Information Worlds

Routledge Studies in Library and Information Science

1. Using the Engineering Literature
Edited by Bonnie A. Osif

2. Museum Informatics
People, Information, and Technology in Museums
Edited by Paul F. Marty and Katherine B. Jones

3. Managing the Transition from Print to Electronic Journals and Resources
A Guide for Library and Information Professionals
Edited by Maria Collins and Patrick Carr

4. The Challenges to Library Learning
Solutions for Librarians
Bruce Massis

5. E-Journals Access and Management
Edited by Wayne Jones

6. Digital Scholarship
Edited by Marta Mestrovic Deyrup

7. Serials Binding
A Simple and Complete Guidebook to Processes
Irma Nicola

8. Information Worlds
Social Context, Technology, and Information Behavior in the Age of the Internet
Paul T. Jaeger and Gary Burnett

Previous titles to appear in Routledge Studies in Library and Information Science include:

Using the Mathematics Literature
Edited by Kristine K. Fowler

Electronic Theses and Dissertations
A Sourcebook for Educators, Students, and Librarians
Edited by Edward A. Fox

Global Librarianship
Edited by Martin A. Kesselman

Using the Financial and Business Literature
Edited by Thomas Slavens

Using the Biological Literature
A Practical Guide
Edited by Diane Schmidt

Using the Agricultural, Environmental, and Food Literature
Edited by Barbara S. Hutchinson

Becoming a Digital Library
Edited by Susan J. Barnes

Guide to the Successful Thesis and Dissertation
A Handbook for Students and Faculty
Edited by James Mauch

Electronic Printing and Publishing
The Document Processing Revolution
Edited by Michael B. Spring

Information Worlds

Social Context, Technology, and Information
Behavior in the Age of the Internet

Paul T. Jaeger and Gary Burnett

Routledge
Taylor & Francis Group
New York   London

First published 2010
by Routledge
711 Third Avenue, New York, NY 10017

Simultaneously published in the UK
by Routledge
2 Park Square, Milton Park, Abingdon, Oxfordshire OX14 4RN

First issued in paperback 2014

Routledge is an imprint of the Taylor & Francis Group, an informa business

© 2010 Taylor & Francis

Typeset in Sabon by IBT Global.

Library of Congress Cataloging-in-Publication Data
Jaeger, Paul T., 1974–
  Information worlds : social context, technology, and information behavior in the age of the Internet / by Paul T. Jaeger and Gary Burnett.
    p. cm.—(Routledge studies in library and information science)
  Includes bibliographical references and index.
  1. Information society.   2. Internet.   I. Burnett, Gary, 1955–   II. Title.
  HM851.J337 2010
  303.48'33—dc22
  2009046103

ISBN13 978-0-415-99778-2 (hbk)
ISBN13 978-1-138-80118-9 (pbk)
ISBN13 978-0-203-85163-0 (ebk)

To my mother, Carol, for her boundless support, and to Mallory and Rosa for keeping me company (and standing on the keyboard, sprawling on papers, and attacking the mouse) while I try to do my work.—Paul

To my wife, Kathy and my children, Joshua—and his wife, Casey—and Jerusha, all of whom put up with me more than I deserve in more ways than I hope to mention. And, particularly, to my grandchildren Matthew and Lindsey; it will be years before you read this, if you ever do, but you still inspire your YeYe.—Gary

Communication systems are neutral. They have neither conscience nor morality; only history. They will broadcast truth or falsehood with equal facility. Man communicating with man poses not a problem of how to say it, but more fundamentally what he is to say.

Edward R. Murrow

You get information from people trying to give you . . . trouble.

Robert Grenier

Contents

About the Authors

Paul T. Jaeger, Ph.D., J.D., (pjaeger@umd.edu) is an Assistant Professor in the College of Information Studies, the Director of the Center for Information Policy and Electronic Government (http://www.cipeg.umd.edu), and the Associate Director of the Center for Library and Information Innovation (http://www.liicenter.org) at the University of Maryland. His research focuses on the ways in which law and public policy shape access to information, particularly in terms of access for underserved populations. Specific areas of interest include information policy, e-government, public libraries and technology, and, of course, social theory of information.

Dr. Jaeger is the Associate Editor of *Library Quarterly*. His research has been funded by the Institute of Museum & Library Services, the National Science Foundation, the Bill & Melinda Gates Foundation, the American Library Association, and the Association of Research Libraries. He is the author of more than eighty journal articles and book chapters, along with six books.

His research has appeared in such journals as *Government Information Quarterly, Library & Information Science Research, Journal of the American Society for Information Science and Technology, Library Quarterly, Information Research, Telecommunications Policy, Journal of Information Technology and Politics, Electronic Journal of Electronic Government,* and *Information Technology and Libraries,* among others. Other recent books by Dr. Jaeger include *Public Libraries and Internet Service Roles: Measuring and Maximizing Internet Services* with Charles R. McClure published in 2008 by ALA Editions and *Public Libraries and the Internet: Roles, Perspectives, and Implications* with John Carlo Bertot and Charles R. McClure published in 2010 by Libraries Unlimited.

Gary Burnett, Ph.D., (gburnett@fsu.edu) is an Associate Professor at the College of Communication and Information of Florida State University, where he has taught since 1996. He holds a Ph.D. in English from Princeton University, where he specialized in modern American poetry, and an M.L.S. from Rutgers University. His research has focused on interpretive practices and the interaction between social interaction and information exchange in text-based online communities, using a variety of approaches and frameworks, including the theory of information worlds and social and textual hermeneutics.

Dr. Burnett is the author of a book on the American poet H.D., and his research has appeared in a number of journals, including *Library Quarterly, Journal of the American Society for Information Science and Technology, Library & Information Science Research, First Monday, Library Trends, Information Research, Journal of the Association of Information Systems* and *Journal of Computer Mediated Communication.*

Acknowledgements

A work of this nature is very long in the gestation, development, and weaving processes. Well before the writing begins, the ineluctably lengthy formulation of ideas has been occurring for some time. In this case, the formulation has been ongoing for seven years. Along the way, many people enriched this process, whether or not they meant to, and even if they yet realize that they have. A number of our colleagues, collaborators, and friends have helped us work through these ideas—either by supporting or by challenging them. Both approaches helped to strengthen the concepts and arguments of the theory of information worlds.

Important folks without whom this book would have turned out differently are: Rebecca Adams, Ben Bederson, Michele Besant, Laurie Bonnici, Harry Buerkle, Kathy Burnett, Christian Crumlish, Allison Druin, Karen Fisher, Ken Fleischmann, Renee Franklin, Crystal Fulton, Jen Golbeck, Michelle Kazmer, Paul Marty, Lesley Langa, Jenny Preece, Mega M Subramaniam, Reijo Savolainen, Kim Thompson, and the kind folks at Deadwood and DSD. We also thank the numerous journal editors and anonymous reviewers who have provided invaluable feedback for our work over the years. Jen Golbeck deserves a great deal of our appreciation (and a bunch of bright red flowers) for providing thoughtful feedback on the completed manuscript. Special mention obviously must also go the late Elfreda Chatman, whose life's work was obviously essential to this project. Both of the authors knew her, and one was fortunate enough to have worked with her (and to still have a collaborative article left unfinished at the time of her death sitting on his hard drive).

We are also very grateful to our editor Laura Stearns for listening to us and deciding to let us write this book, to the anonymous reviewers of the book proposal who offered many strong suggestions, and to Routledge for publishing it. As long time admirers of the theoretical and other works that have been published by Routledge through its many years, we are delighted to have our information theory book appearing through this storied imprint.

Finally, we would like to express our appreciation of the readers of this book. Thank you for spending time with our ideas.

Introduction

The theory of information worlds—created to provide a framework by which to simultaneously examine information behavior at both the immediate and broader social levels—is the result of trying to bridge the canyonesque gaps between the ways information is viewed in terms of small social units and the ways it is viewed in larger societal and political processes. Since information and accompanying information technologies underlie virtually every aspect of life in technologically advanced societies, the failure thus far to make stronger theoretical connections between information behavior in the various levels of society is utterly perplexing.

Part of the explanation likely results from the difficulty and complexity of working with the theoretical dimensions of information, particularly social aspects such as information behavior. Regardless of the reasons, there can be no quivering about the importance of information theory. It's not that we are lacking good people working on these issues; the problem is that there are not enough people working on these issues. The slender amount of work in this area has very serious consequences for the ability to research and understand the true roles that information plays in personal, social, and political arenas. As scholars of information issues and information professionals, it behooves us to work toward understanding the totality of the complex and interrelated roles of information as it moves between social units, affecting social interactions at the smallest and largest levels.

This book argues that the theory of information worlds can serve as a theoretical driver both in library and information science—the authors' native discipline—and across fields, aiming to enrich and expand our conceptions and understanding of the multi-layered role of information in society. It has been written to have implications in a variety of domains related to information use and provision, including information behavior in context, library services, information seeking, communication, information technology, and policy. As part of this intention, the ideas presented herein can serve as a framework for advances not only in theoretical work in library and information science and other fields but also for empirical investigations of the social dimensions of information and information use.

While this book may most naturally be considered a library and information science text, the ideas have applications for researchers in the many fields that deal with information as a foundational element, including communication, computer science, education, human-computer interaction, media studies, public policy, and political science, among others. This work has liberally incorporated research from these fields, which has hopefully both strengthened the arguments and increased the utility of the ideas it presents. One of the key goals of this book is to encourage discourse and collaboration across fields in tackling the most complicated problems of information behavior in society.

To encourage such dialogue, the book has been written from a perspective that is not tied to a particular society. While many of its examples are drawn from the United States, where both of the authors live and work, the book uses cross-cultural examples and issues pulled from nations and events around the world. The issues discussed are of relevance to all societies facing the challenge of understanding the social aspects of information flows and behavior in the electronic environment.

The issues presented are also, to varying extents, of an inherently political nature. While some scholars resist dealing with current political events in an effort to avoid seeming partisan, such an approach would be contrary to the purpose of an endeavor such as this one. The social groups in power and the ways in which they use their power have a significant impact on the role of information in a society. Decades of partisanship and intense polarization in the United States now allow any argument that is perceived by a political faction as negatively portraying them to be disingenuously attacked as partisan even if it is simply detailing recent events. The discussions of political issues and the role of information in political processes are not intended to be an endorsement of one party or another. We live in a political world, as Bob Dylan has noted, and the political issues are simply a part of the story that must be told.

The book has been written to reach those with a number of interests, across fields and countries. Key themes include (1) the need for better theoretical understandings of information behavior in society; (2) the increasing importance of social information behavior as a central research problem in a growing range of fields; (3) the need to bridge a broad number of streams of theory and research about information in society; and (4) the links of theory to important information contexts such as online communities, libraries, media, public policy, and the political process.

We hope that some of our readers will be encouraged to join the discourse devoted to developing and refining social information theory. There are far too few studies and explorations of theory related to information behavior, particularly those linking information behavior to social and political contexts. And joining the discourse on information theory should not seem like an unachievable goal for any scholar interested in issues of information. One of the authors (Jaeger) was initially surprised to find himself working on theoretical concepts, having come to work on theory building in response to the general lack of consideration given to information in the theories of political science and public policy. The other author (Burnett) has made information theory development a key part of his career and even collaborated with one of the theorists whose work is foundational to this text.

The ideas in this book are the result of seven years of collaboration between the authors, being slowly formulated across a series of journal articles and conference presentations (Burnett & Jaeger, 2005, 2008; Burnett, Jaeger, & Thompson, 2008; Jaeger & Burnett, 2003, 2005). Resulting from initial explorations of the intersections of concepts of small worlds and information behavior related to public policy, the line of thinking evolved into trying to explore the linkages between individual, small group, community, and society-wide information behavior driven by social forces large and small.

At some point, it became clear that theory development was a key part of what was occurring in the work, though the naming of the theory took much longer than the admission that theory development was afoot. The first complete articulation of all of the concepts and components of the theory can be found in Burnett and Jaeger's 2008 paper "Small Worlds, Lifeworlds, and Information: The Ramifications of the Information Behaviors of Social Groups in Public Policy and the Public Sphere." The only structural element missing from that paper is the titular designation of theory of information worlds. While this book provides a considerably detailed expansion and elaboration of the theory, that article provides the first complete articulation of theory. Based on the author order of that paper, it also means that the theory itself would most properly be referred to as Burnett and Jaeger's theory of information worlds.

Ultimately, the core mission of this book is the presentation of the theory of information worlds and its potential applications for research, teaching, and practice related to information in society. The authors of this book, not surprisingly, feel that the ideas herein have a goodly amount of potential to advance thinking about information in society.

Paul T. Jaeger and Gary Burnett
June 2009

1   Theory, Information, and Society

Theoretical frameworks for understanding information within a social context are frustratingly rare. Even though information is central to any developed or developing nation—in fact, the loss of technologies to transmit information would be as catastrophic as the loss of technologies to transmit electricity or water at this point—theory addressing the social roles and impacts of information is unfortunately hard to come by. However, information is the true driver of interpersonal interactions, civic engagement, business operations, political discourse, and every other physical and virtual interaction in an age defined by the omnipresence of information and communication technologies (ICTs). Since the first conceptualizations of democracy, information access and exchange have been seen as its most essential foundations. And if Foucault (1979) is correct that reason cannot act as a tool of oppression, then information not only feeds democratic discourse, it also fuels the reason that limits oppression. Theoretical frameworks, to truly capture the depth and breadth of the roles of information in society, need to capture all of these aspects of information.

The difficulty of capturing the wide range of social roles of information, however, does not alone explain the paucity of theory of the social contexts of information. It has been aptly noted that "social theory is challenged in the information age" (Fuchs, 2008, p. ix). Information is hard to capture as a concept; it changes forms and transmission methods and permeates society in uncountable ways. The vast and unblinking developments of ICTs mean that the channels of information evolve at an amazing pace. While ICTs can both create chaos and "partly help to reconstruct an order" (Gitelman, 2006, p. 155), the relentless pace of changes in ICTs may now resist the reconstruction of order. Ironically, historical predictions about the future of technology usually revolved around vaporware—technologies that are always on the way, but never quite materialize (Duguid, 1996). In contrast, the evolution of technology now outpaces the imaginations of many of the people who study the future of technology. For example, in 2002, Lawrence Lessig, in a forecast that clearly turned out to be less than accurate, predicted that "AOL Time Warner and Microsoft—[would] define the next five years of the Internet's life" (p. 267).

With policies related to information emanating from local governments, state and provincial governments, national governments, and supranational and non-governmental organizations, the information policy environment is more complex and complicated than at any previous point in history (McClure & Jaeger, 2008a; Relyea, 2008). As information and ICTs have mushroomed in importance in interpersonal, financial, educational, professional, and governmental transactions and interactions with the rise of the World Wide Web, many new information policies have emerged. The legal landscape "has become filled with laws and regulations dealing with information and communication," as more than "600 bills dealing with the Internet alone were on the table during the 107th Congress" in the United States (Braman, 2004, p. 153). Information policies both address a societal issue regarding information and attempt to balance the interests of different stakeholder groups impacted by an issue (Thompson, McClure, & Jaeger, 2003). New ICTs often compel governments to alter policies to fit the new technical environment, but the policy activity related to information is not always meant to promote growth and adoption of new technologies related to information. The profusion of policy around the globe has closely paralleled the enormous changes in ICTs over the past fifty years.

New ICTs (e.g., the printing press, telegraph, radio, television, railroad, and telephone) have long influenced the political processes and the functioning of governments (Bimber, 2003). Their rapid development creates questions for the longest traditions; for nearly a millennium, law in the West has been conceived of as a constantly growing and evolving body of concepts that change to meet the social and technological changes over time (Berman, 1983, 2003). Yet law simply cannot be conceived, debated, and passed at a pace to keep up with current levels of technological change (Braman, 2006; Grimes, Jaeger, & Fleischmann, 2008; Jaeger, Lin, & Grimes, 2008; Jaeger, Lin, Grimes, & Simmons, 2009).

Part of this problem is that the tidal wave of social, political, and technological changes resulting from the revolution of the Internet and other ICTs "is understood primarily as a *technical* one" (Boyle, 1996, p. ix). The focus on the technical elements—by designers, scholars, lawmakers, and citizens—often obscures the information issues. These assumptions are evidenced by the disjunction between expectations for and the actual use of technologies. The social, commercial, and policy expectations of technological innovations are usually misplaced, with new technologies driving change in unexpected ways (Brynin, Anderson, & Raban, 2007). The disjunctions likely have a significant relation to the information dimensions of the ICTs.

These disjunctions are further born out in the fact that the mere presence of a new ICT does not necessarily change quality of life measures in many users. Statistically, the uptake and usage of ICTs makes little difference in the quality of life for users, particularly in terms of direct impacts on everyday activities (Anderson, 2007). "Despite the large number of policy references to the positive effects of ICTs on people's lives, few of these claims are supported by empirical research" (Heres & Thomas, 2007, p. 176). Studies such as these focus on the ICTs, not the specific types of information they are used to access. And, yet, there is no reason to assume that access to new ICTs necessarily leads to greater information access and usage related to important issues.

ICTs have created the means for creating and sharing information at previously unthinkable levels, resulting in information overload if one tries to conceive of the amount of information now available. One of the most common reactions to information overload is to simply ignore the amount of information available, which seems to be a widespread reaction among both members of the public and scholars (Goulding, 2001; Wilson, 1996). Studies in sociology and psychology indicate that individuals now experience culture as a series of fragments of information, even though people are more likely to recall information correctly and efficiently if it fits within their established cultural frameworks (DiMaggio, 1997; Martin, 1992).

Problems of overload do not even touch on the problems of quality of information or the problems of polarization. A longstanding, perhaps overstated, and yet still largely unaddressed concern about online information is the inaccuracy of much of that information and the ways to help educate citizens about assessing the quality of information they encounter. Group polarization, in contrast, describes the situation where people only seek out information sources coming from people whom they perceive as like them and that validate and reinforce their already held beliefs and opinions, an activity that the Internet can make easier (Jaeger, 2005). To put it bluntly, "a simple correlation between the Internet environment and the expansion of global civil society can no longer be taken for granted" (Deibert & Rohozinski, 2008, p. 146). Nor can the reverse—that the Internet is a harbinger of the collapse of civil society—be assumed. As is argued elsewhere in this book, the Internet is perhaps most accurately seen as a kind of "both/and" setting for information, simultaneously offering the potential for new avenues of information access and social participation and bringing a risk of further polarization and misdirection. All of these strands are woven together into an extremely complex and daunting tapestry against which scholars must try to examine the roles of information in society.

The greatest barrier to the development of theoretical frameworks for information in society, though, may be in the ways that academic disciplines have conceived of the study of information. As people begin to experience rapid social transformations on a day-to-day basis, social sciences have a difficult time keeping pace as basic definitions and assumptions change quickly (Beck, 2002). At a deeper level, however, information as a theoretical concept does not necessarily receive the attention it deserves. While information is vital to and underlies every academic discipline, it is usually left to library and information science (LIS) schools to truly focus on it. This is highly problematic for two reasons. First, the foundations of

LIS are squarely in professional education, making many of the faculty in LIS reluctant to engage in theoretical research. As a result, though there are some notable expectations among its researchers, LIS is relatively lacking in native theory (which is discussed in detail in Chapter 10).

The second problem is grounded in the fact that many other fields tend to ignore the links between their theoretical work and issues related to information, resulting in the centrality of information to their studies being insufficiently recognized, as in much of the work drawing upon Shannon and Weaver's (1964) theory of communication, which largely treats information as a simple and unambiguous signal passing through the conduit of communication systems. When information is studied in other fields, far too often it is conflated with ICTs, as if the issues of content and method of transmission were interchangeable. For example, many major works about information and the political process limit their focus to ICTs (e.g., Barber, 1994; Dahl, 1989; Davis, 1998; Davis & Owen, 1998; Etzioni, 1993; Wilhelm, 2000). Yet, one rare exception to this trend that focuses on information and the political process makes no connections to, or even mentions the existence of, research from LIS (Bimber, 2003).

These problems are also part of a long-recognized issue that different disciplines of social science do not communicate well, often leading to terminological and methodological confusion (Bain, 1943). In spite of these challenges—both old and new—social science must develop meaningful frameworks for studying and understanding the social contexts of information. As was noted three decades ago in a paper about LIS' struggles with theory, "there is nothing worth noticing until you have a theory . . . One must start with a guess, or theory, then collect data in the light of the guess to see what it reveals" (Swanson, 1980, p. 78). Without such frameworks, social science will be pulled further and further from the actual experiences of the members of society. In a life henged round by information and ICTs, the accurate study of society can only be achieved when the roles of information and ICTs are central to scholarship and to professional practice that is built on the findings of scholarship.

AN OVERVIEW OF THE THEORY OF INFORMATION WORLDS

While the remainder of this chapter and Chapter 2 will work to lay out the components of the theory of information worlds and the origins of these components, the basic structure is presented here as a roadmap. The goal of the theory of information worlds is to enhance our understanding of the role of information in society by providing a means by which to analyze and understand the myriad interactions between information, information behavior, and the many different social contexts within which they exist.

The theory asserts that information behavior is simultaneously shaped by immediate influences, such as friends, family, co-workers, and trusted information sources of the small worlds in which individuals live, as well as larger social influences, including public sphere institutions, media, technology, and politics. The framework of this theory can be used to examine the contexts of information at the micro (small worlds), meso (intermediate), and macro (the lifeworld) levels of society. These levels, though separate, are intimately interrelated.

Though the theory of information worlds draws upon work from a wide range of disciplines and ties together elements of many social theories, the largest contributors to the foundation of the theory are the theoretical works of Jürgen Habermas and Elfreda Chatman. Habermas was interested in the largest social structures, while Chatman was most interested in the smallest social units. In contrast, the theory of information worlds explores information behavior in terms of all of the intertwined levels of society—the small worlds of everyday life, the mediating social institutions, the concerns of an entire society, and the political and economic forces that shape society—which are constantly shaping, interacting, and reshaping one another.

To examine these levels and structures in society, the theory of information worlds focuses on five social elements that are part of every level of society:

- Social norms, a world's shared sense of the appropriateness of social appearances and observable behaviors
- Social types, the roles that define actors and how they are perceived within a world
- Information value, a world's shared sense of a scale of the importance of information
- Information behavior, the full range of behaviors and activities related to information available to members of a world
- Boundaries, the places at which information worlds come into contact with each other and across which communication and information exchange can—but may or may not—take place

These elements are interrelated and constantly interact with and influence each other.

At the micro level, each small world is a social group with its own social norms, social types, acceptable forms of information behavior, and shared perceptions of information value. Within a given small world, members develop normative ways in which information is accessed, understood, and exchanged both within the small world and with others outside that world. Individuals typically exist in many small worlds—such as friends, family, co-workers, and people with shared hobbies—and individuals will generally conform to the norms and expectations of each small world when they interact with other members of that world. Each world has many places where its members might interact with members of other small worlds, and these points of contact serve as the boundaries between different worlds. The number of small worlds is by no means static, as contact between small worlds and other inputs from social structures can lead to the creation of new small worlds and the disappearance of existing ones.

Information moves through the boundaries between worlds via people who cross between the different worlds to which they belong and through interactions between members of multiple small worlds in physical and virtual social spaces where members of different worlds encounter each other. Exposure to the perspectives of other worlds occurs through physical public sphere institutions, such as public libraries or schools, and through new technological avenues of communication and exchange, such as social networks on the Internet. As information moves through the boundaries between worlds, the social norms of each world shape the ways in which that information is treated, understood, and used, creating different roles for the information within each world.

As a larger collective, these small worlds constitute the lifeworld of information—the full ensemble of communication and information exchange in a society. The various perspectives of the small worlds, as they come into contact with one another, will influence the overall place of information within the broader lifeworld. The perspectives of the small worlds, however, are not the only influence on the place of information in the lifeworld. Other influences, originating in the small but powerful worlds of the media, the marketplace, or the government, can either promote the movement of information between small worlds or constrain such movement, constricting the socially acceptable perceptions of information. Certain public sphere institutions—such as libraries and schools—exist specifically to ensure that information moves between the small worlds and that members of each small world are exposed to the perspectives of many other worlds.

Many of the influences on small worlds and the lifeworld are inherently neutral, capable of advancing goals of either increasing or decreasing information access and exchange. Through time, many different types of ICTs have both acted as vehicles by which information worlds have connected and interacted in new ways and as tools used by powerful information worlds to constrain or limit the flow of information through and across worlds. Currently, the Internet and online social networks may be the most powerful examples of this dual role of ICTs. Small worlds are shaped by all of these larger worlds and forces, but also, in turn, exert their own influence on them.

Building upon all of these foundations, the theory of information worlds is designed to account for all of the elements at work in shaping the role that information plays within a society. The entire number of small worlds and the lifeworld of a society are an information world in the most expansive sense. However, many smaller and intermediate information worlds also exist in a society as groups of small worlds that are bound together in some familial, community, social, professional, educational, cultural, political, geographical, technological, or other means create other units with an interrelated set of approaches to information.

Though by necessity much more complex than approaches to studying information at the micro, the meso, or the macro level, the theory of information worlds offers a much richer and nuanced understanding of the ways in which information is perceived and moves though society. There is no intention to undermine the great benefit and value of studying the levels individually, but the large-scale perspective has not previously been sufficiently explored in social theory of information. The theory of information worlds, thus, provides the researcher another approach and related conceptual tools, which can be used to create a thorough and realistic picture of information across society.

INFORMATION WORLDS: TOWARD A CROSS-DISCIPLINARY THEORY OF INFORMATION

One of the most promising ways to develop the necessary theoretical frameworks for information is to draw simultaneously from both LIS and other traditions. The authors of this book have been working together for years to build a theory that combines the best elements of LIS for understanding the social impacts of information and of other social sciences for understanding the concurrent impacts of society upon roles and expectations for information (e.g., Burnett & Jaeger, 2008; Burnett, Jaeger, & Thompson, 2008; Jaeger & Burnett, 2003, 2005). Drawing on the concepts from two theoretical precursors (philosopher Jürgen Habermas and information theorist Elfreda Chatman), the theory detailed in this book represents the melding of these concepts into a single theoretical model which has been designated the theory of information worlds. The theories of these scholars examine the ways in which information is embedded in the social worlds of people, but they do so from two very different perspectives: Chatman focuses almost solely on the place of information in very specific localized communities, while Habermas examines information strictly in terms of the sum total of information and communication resources of a society as a whole. However, Chatman largely ignores both the broader "lifeworld" within which her "small worlds" exist and situations in which multiple small worlds come into contact (or conflict) with one another, while Habermas pays little if any attention to the ways in which the broad lifeworld might interact with or be realized in localized contexts and specific communities.

The foundational element of Habermas' work is the concept of the public sphere, defined as "the sphere of private people come together as a public." Integral to the concept of the public sphere is what he called the lifeworld, a broadly defined collective environment of information and communication that links members of an otherwise disparate society together. Habermas conceived of the lifeworld as a macro-level concept that comprised the sum total of information and communication resources and activities of a society. His work pays little attention to the ways in which such a broad concept might be realized at the local level, within specific communities.

In contrast, Chatman developed a series of small world concepts, culminating in her theory of normative behavior and providing a conceptual framework for understanding the place of information within specific, small, localized, and largely homogeneous communities that she called small worlds. Chatman's work conceptualizes information and information behavior not as a macro-level or culture-wide phenomenon, but at the micro level of specific communities. However, her work pays little attention either to the broader lifeworld within which such small worlds exist or to situations in which multiple small worlds come into contact with one another, often with conflicting sets of social norms and conflicting understandings of the role of information.

Combining the concepts of Chatman and Habermas into a multi-level theory of information worlds allows a richer understanding of the intersections between information and the many different cultural contexts within which it is used, from the macro to the micro. Drawing upon and greatly extending previous work by the authors, this book uses the theory of information worlds to examine a variety of issues, including information access and exchange, the varying values assigned to information by different social groups, information policy, information provision and libraries, physical and virtual means of communication, the role of information in a democracy, and technological change. Given the vital concerns of the theoretical study of the social roles of information, Habermas and Chatman present an ideal combination of elements about the place of information both at the macro levels of society as a whole and at the micro level of interpersonal interactions and daily life.

Habermas' conception of the public sphere stands as a preeminent example of the importance of information to the functioning of a democratic society. Though early inklings of the concept of the public sphere can be found in the work of Immanuel Kant, C. Wright Mills, and Hannah Arendt, the works of Jürgen Habermas truly defined the notion, particularly 1981's *The Structural Transformation of the Public Sphere: An Inquiry into Category of Bourgeois Society* (Gilman-Opalsky, 2008). The public sphere evolved from appeals to the public in an attempt to develop personalized links, private testimony, and individual power into public, impartial, and disinterested forms (Habermas, 1981). Ideally, the public sphere is an arena of common problems—social, economic, and political—that are considered as a conversation of equals through multitudes of perspectives, interactions, and communications through publicly available channels, with arguments based on reason and the better argument (Alejandro, 1993; Clark, 2000; Giddens, 1985; Green, 2001). As a result, Habermas is "the key ideologist of a particular and influential take on human rights" (Bowring, 2008, p. 104).

As a means to understand the importance of information to democracy, the strength of the public sphere "lies in the presumption of reason, the human ability to define and solve problems" (Boeder, 2005, n. p.).

Habermas' conception, however, only succeeds at the highest social levels. Chatman's work offers a window into the purely local and narrowly contextual settings of information by focusing on the values and perspectives toward information of individual social groups (Burnett, Besant, & Chatman, 2001). The small world is a social group in which "mutual opinions and concerns are reflected by its members" (Chatman, 1999, p. 213). Within each small world, everyday activities are considered to be "the way things are" and are frequently taken for granted as being standard to all small worlds even when they are not. Chatman's work supports a close analysis of situations in which different small worlds intersect, leading to conflicts and misunderstandings between groups (Burnett, Jaeger, & Thompson, 2008). The direct, real world sources of Chatman's theories also serve as a balance against the criticism most commonly leveled against Habermas' theories—that they are too idealized and too far removed from the messy and nitty-gritty realities of peoples' day-to-day lives.

The importance of information moving both through and between social groups has long been recognized. For decades, psychologists and sociologists have studied the importance of groups in exchanging information and shaping opinion. It was first documented in 1935 that the perspectives of people with divergent separate judgments will move closer to each other significantly when exposed to one another: "when two or three individuals give their judgments in the presence of each other, the whole group establishes a range and a point of reference particular to the group" (Sherif, 1935, p. 52). Further, being a part of group gives individuals a stake in the context and actions of the group, since a person, "as soon as [he] is in the midst of a group . . . is no longer indifferent to it" (Asch, 1952, p. 485). The exchange of information between social groups serves to increase the number of connections between them, the range of information available to their members, and the amount of social capital (Coleman, 1988; Granovetter, 1973). "Social capital arises from relationships between individuals embedded within a group, and is developed, maintained, or dispersed as a result of interpersonal and group processes" (Ling, 2007, p. 151). Such social capital is clearly not possible without information flows between groups.

The ability to understand the concurrent micro, meso (or mid-level; see DiMaggio, 1997; Powell & DiMaggio, 1991), and macro flows of information is essential to a theory of information in society as the levels are inextricably interconnected. Though members of many social groups may prefer interpersonal communication, "you cannot have face-to-face relations with ten million or a million or a hundred thousand people" (Appiah, 2005, pp. 216–217). Simultaneously, information exchanged within social groups cannot be understood without also considering the greater social context in a society defined by information and ICTs. Social forces not only influence how individuals in specific social settings conceive of and use information but also strongly influence how decision makers and information professionals conceptualize information and access in the first place. As such, theories must be developed that help to understand these two linked levels of information flows. Further, although a considerable amount of theoretical work examines information flows at the scale of the global economy and through the infrastructure and telecommunications systems that make such an economy possible (see, for instance, Castells, 2000), there is a clear need for theoretical approaches to information flows and activities at levels other than the global.

This understanding of the interconnected micro and macro information flows is acutely important given the sizeable differences between the idealized public sphere and the current health of the public sphere. In one sense, social groups will always struggle for their perspectives and beliefs to have preeminence in society (Williams, 1958, 1968). However, many scholars, including Habermas, have noted the long-term decline of the health of the public sphere as corporate and government influence have overrun many channels of access to and exchange of information in the public sphere (e.g., Clark, 2000; Dahlgren, 1995; Habermas, 1996a). In this context, the insights provided by Chatman's work provide vital understandings of information in the social groups that are not hegemonic elements of society.

In a sense, the goals of developing the theory of information worlds parallel the goals that have driven the development of other theories that have sought to bridge different levels of a particular phenomenon or system. For example, Anthony Giddens (1991), in describing the development of his theory of structuration, has explained that his work was intended to provide a means to understand "how actors are at the same time the creators of social systems yet created by them" (p. 204). The theory of information worlds has been developed to bring a similar sense of balance to the study of the social roles and impacts of information throughout the separate yet intertwined micro, meso, and macro levels of society.

A core foundational element to this exploration of information in society is a starting point for understanding the meaning of "information." While entire books have been devoted to defining it (e.g., Borgmann, 1999), information is often discussed with no attempt at a definition. It is entirely possible, for example, to write a book on information organization without offering a definition of information (e.g., Taylor, 1999). When it is defined, it can be understood to be many different things. Information can be explained in terms of its attributes (Svenonius, 2000) or in terms of its engineering aspects rather than its content (Shannon & Weaver, 1964). Information can be divided in categories of process, knowledge, and object (Buckland, 1991a, 1991b). Information can also be defined by what it is not, such as Brown and Duguid (2002) defining it entirely by explaining how "knowledge" is superior to "information" or Buckland's (1997) differentiation between information and a document that contains it. In fact, the word "information" carries numerous definitions and contexts, with technical uses and common uses having become intertwined and conflated in discussions of the information age (Nunberg, 1996).

In this book, information should be understood very broadly. It can be seen as the sum total of the content—facts, knowledge, feeling, opinions, symbols, and context—conveyed through communication between individuals or groups through any physical or virtual medium. Much of this book focuses on information about broadly significant and pressing social and political issues. This focus, however, is not meant to imply that other kinds of information are irrelevant or trivial, or that information about mundane concerns, pop culture phenomena, or personal passions does not play an important and meaningful role in people's lives. Similarly, it is important to remember that people, whatever kind of information they find of value, interact with information in a wide variety of ways. As Wilson (2000, p. 49) put it, this theory is related to "the totality of human behavior in relation to sources and channels, including both passive and active information-seeking, and information use," just as the theory relates to information of all types.

Still, while a specific individual may find great value in information that, to others, may appear trivial, the focus of the public sphere on information related to governance drives much of the discussion herein. Valid questions can be raised about the strength of connections between information resources and democratic health when considering the individual citizen (Dervin, 1994). However, in a collective sense, the ability to discuss the actions of a pop celebrity or the plot permutations of a television serial and the ability to discuss the actions of a president carry very different import to the overall health of a public sphere and its conjoined democratic society.

THE STRUCTURE AND ARGUMENT OF THE BOOK

Building upon the issues discussed previously, Chapter 2 establishes the primary theoretical framework for the book. This chapter argues in great detail that combining elements of the theories of Habermas and Chatman into a multi-level concept of information worlds allows a richer understanding of the intersections between information and the many different cultural contexts within which information exists, from the macro to the micro, and the resulting information behavior. This chapter introduces the ways in which these worlds manifest and shape a number of different contexts, including not only institutions and ICTs that are dedicated to making information available but also more amorphous worlds in which the exchange of information is embedded in day-to-day life and social interaction.

Chapter 3 explores the central question of value assigned to information by different social groups and individual perspectives related to their information worlds. Within society, there is a sizeable tension tied to two key issues: beliefs regarding information value and—growing out of that— beliefs regarding who should have access to what types of information. These beliefs likely will change relative to the importance and complexity of the information at hand, but they are always defined by social context. In terms of information about meaningful issues, such as social, ethical, and political questions, perceptions about the appropriate value of and appropriate forms of access to information will shape views about levels of openness, secrecy, and control that are acceptable in society and in individual lives.

Divergent views on these issues present a continuum of perceptions that highly influence the access, exchange, and use of information in personal, social, corporate, and governmental contexts. This chapter explores these issues through lenses such as social tagging and intellectual property. The attitudes range from seeing information as a mechanism of control and surveillance to seeing it as vital to an informed citizenry through democratic dialogue, education, and open access. Historically, governments—even democratic governments—have generally favored focusing on information as a mechanism of social control, legal authority, and surveillance to retain power. Similarly, media also often use information as a means of control over public discourse. Citizens and public sphere entities, however, often actively favor greater levels of access and exchange within information worlds. While there are very clear contexts in which limits on information access are essential (such as protecting intellectual property and protecting government secrets), too much control of information can stifle open access and exchange, the quality of information available, education, and even democracy itself.

Ultimately, information value can have tremendously important social, economic, political, and ethical implications. The ways in which information can move through information worlds hinge on the values that society—and its constituent worlds—gives to it. Democratic governance is sustainable only when the government and citizenry are sufficiently committed to and vigilant in preserving the values of open access and exchange. Great changes in the ways in which information is valued will lead to significant changes in its use by governments, corporations, and individuals.

Chapter 4 focuses on the implications of the theory of information worlds for access to and exchange of information. Access to information can be conceptualized in many ways; different academic disciplines, including LIS, have approached it in terms of knowledge, technology, communication, control, commodities, and participation, with influences on access including physical, cognitive, affective, economic, social, and political issues. LIS scholars have tended, however, to approach information access in two specific ways: as a matter of physical access, emphasizing the format and location of documents, and the degree to which users can physically acquire and use those documents and as a matter of intellectual access, emphasizing users' cognitive abilities as well as mechanisms and structures for organizing and categorizing information. This chapter will examine the historical roots of these conceptualizations and will discuss current systems that are in place to ensure that access to information is possible.

Earlier work by the authors has proposed a third dimension for understanding access to information: social access. One important element of the concept of information worlds (as of Habermas' lifeworld and Chatman's small worlds) is that they are built around shared understandings of and behavioral norms regarding information. Not only is information exchanged by members of a given small world as part of their ongoing social interaction, but multiple small worlds (as component parts of the larger lifeworld) intersect with each other in terms of the ways in which they understand and use information. This chapter presents information access in terms of such social aspects and will argue that information access and exchange are best understood in terms of the small- and large-scale social contexts within which they take place.

Chapter 5 bridges the theoretical foundations of the book and the applications of these theories in various social contexts by tracing the evolution of the public sphere as physical and virtual place through the lens of the evolution of the social roles of libraries. As longstanding pillars of the public sphere, libraries through the past century and a half serve as an ideal case study of the changes in the nature of public sphere entities. As many other physical public sphere entities—such as the archetypical village green—have largely vanished, the library became the only type of physical public sphere entity generally found in most communities. This chapter explores this evolution in relation to what the public sphere entities have viewed as their responsibilities, the social roles and expectations for these entities, the roles of technology in the public sphere, and the migration from the public sphere as a purely physical place to a frequently virtual place.

This examination focuses on these changes not only from the perspective of the citizen but also from the perspective of the individuals who work in public sphere entities. In less than a century, librarians moved from viewing their professional mission as acting as moral arbiters to the dramatically different mission of defending free access to all kinds of information. In the same period, libraries have evolved from physical entities where many different small worlds met and interacted to places where patrons engage in more and more public sphere activities in geographically dispersed small worlds through the computers and Internet connections provided by libraries. While libraries still serve as hubs of public sphere engagement, spreading information throughout small worlds, many librarians are currently resisting the prominence of technology in the social roles of public libraries. All of these changes in libraries, librarianship, and technology serve to illustrate the interrelated evolutions of physical and virtual public spheres and their changing impacts on information worlds.

As discussed in Chapter 6, both traditional and newer technologies have an important influence not only on what information is available in information worlds (large and small), but also on how that information is disseminated, preserved, and used. This chapter investigates historical and current trends in the use of information technologies, examining how those technologies impact the availability and movement of information in information worlds. The primary focus of the chapter is on several recent technologically rooted phenomena that meld social interaction with information exchange. The information-related activities—in online communities, in blogs and wikis, as well as e-government—present myriad issues regarding the ways in which the Internet may be shaping the future of information access and exchange.

Such a melding of social activity and information exchange, in the context of technologically mediated interaction, has begun to transform information worlds. Phenomena like blogs and wikis have clearly become significant parts of the current overall information lifeworld, enhancing the ability of individuals and specific narrowly focused small worlds to access information and to become information providers themselves without requiring them to adapt themselves to the controls and rigors of traditional publication media. At the same time, however, increased reliance on access to the technological tools needed to sustain these activities may have important consequences for many small worlds that either distrust the technologies or simply do not have reliable access to them. It is of increasing importance that ICTs be deployed in such a way as to enhance information access and exchange for all instead of widening the gaps between those with ready access and those without.

Chapter 7 examines the ways in which both the amount and quality of information available for discourse about social and political issues at all levels are heavily dependant on the mass media in a society. Though the formats and methods of information dissemination have evolved, their importance led Habermas to suggest the public sphere itself was impossible without an active and vigorous media. In democratic societies, the media has played many different roles, but, regardless of the nature of such roles, it has long been considered a vital part of an informed democratic population. The content of media—mainstream and non-traditional—heavily influence the discourse by making information available, by presenting it in a certain manner, and by not presenting other information at all.

Within a society, different small worlds have varying relationships with the media—some reject it entirely in favor of the opinions of people they know, some consume it without thought, and most only pay attention to outlets that suit their personal views. However, with growing corporate and government influence over media content over the past several decades, the views available in print and broadcast media have actually constricted and become homogenized as sources of information across all types of information worlds. While the online environment offers many new sources of information and the means of creating new small worlds through which individuals can exchange information, it also presents avenues by which small worlds can become more insulated from the larger information world and can gravitate toward more extreme positions.

As independent media presenting varying perspectives on important social and political issues has long been held necessary for a functional democracy, the historical development and current state of media is core to understanding information behavior in small worlds and in societies.

Chapter 8 explores the relationships between politics and information worlds. Since the terror attacks of 9/11, the U.S. government has employed public policy and political pressure to implement a radical series of changes to the amount of information available to the public, the ways in which the public can access and exchange information, and the amount of information that the government can collect about individuals. While the Bush administration demonstrated a strong inclination to control information even prior to 9/11, the "war on terror" was used by the administration as a means of reshaping information in society. The Bush administration, then, presents a case study in how a government can try to use policy, politics, censorship, intimidation, the legal system, and new technologies to influence how information flows among information worlds (large and small) in a democratic society.

Any government that engages in information politics works to make the uses of information in society conform to its own perceptions of information value, which serves to further alter the meaning of information access and exchange in a society. Polices that limit information access serve multiple purposes, accomplishing both policy goals and partisan political goals. The perspective of the Bush administration appears to have been based on a belief that access to information, in general, should be very limited and tightly controlled by the executive branch of the federal government. It also appears to include the belief that information that is made available for access should fit with the administration's own beliefs and norms. Taken together, these perspectives had significant impacts on the amount and types of information available for access and presented serious questions about the long-term impact on democracy in the United States and internationally. The information policies of the Bush administration cannot be separated from their practical political goals, such as limiting public discussion of administration actions, protecting private interests, and increasing public fear. All of these uses of information politics challenged the core ideals of democracy, a concept of government that is based on the premise of an informed citizenry. Further, many other democratic governments around the world have followed the lead of the Bush administration on many of these issues.

As a potential counterbalance to the information politics of the Bush administration, certain public sphere entities and small worlds have defended the principles of information access. Traditional entities like libraries have worked aggressively to resist the information controls and information collection programs that the Bush administration implemented, while newer entities of the expanded landscape of technologically supported information worlds have worked to provide free access and exchange through online information channels like e-mail, websites, wikis, and blogs. These entities have significant amounts of social capital and are strongly trusted by their members and communities. However, their political power is fairly limited when compared to a federal government. The information value that ultimately takes prominence will significantly shape the ways in which information functions in democratic societies, which in turn will determine how democratic societies really can be.

Chapter 9 wraps together the ideas introduced in the book and examines possible future directions of these concepts. A key part of this chapter is the analysis of the conceptual and practical research questions and opportunities raised by the concept of information worlds for LIS and other disciplines. Finally, Chapter 10 concludes the book with a meditation on the importance of social theory to the future study of information and ICTs.

This book ultimately argues that the theory of information worlds can serve as a theoretical driver both in LIS and across disciplines, aiming to enrich and expand the understanding of the multi-layered role of information in society. It follows the observation that "critical social science is concerned not simply to predict, nor indeed to comprehend. Rather, it intends to transform" (Marks, 2000, p. 129). The ideas discussed in this book have important implications for scholars in a variety of domains related to information use and provision, providing a framework for theoretical and empirical investigations of the social dimensions of information, information behaviors, and ICTs.

2 Information Worlds

The theory of information worlds grows out of concepts drawn from LIS scholar and theorist Elfreda Chatman (small worlds) and philosopher Jürgen Habermas (lifeworld). The theories of these scholars both examine the ways in which information is embedded in the social worlds of people, but they do so from two very different perspectives. Chatman focuses almost solely on the place of information in very specific localized communities, while Habermas examines information strictly in terms of the sum total of information and communication resources of a society as a whole. However, Chatman largely ignores both the broader societal context within which her communities exist and situations in which multiple communities come into contact (or conflict) with one another, while Habermas pays little if any attention to the ways in which the broader society might interact with or be realized in localized contexts and specific communities (Burnett & Jaeger, 2008).

Bridging and extending the concepts of Chatman and Habermas into a multi-level theory of information worlds provides a deeper and more nuanced understanding of the intersections between information, information behavior, and the many different cultural contexts within which information is used, from the macro to the micro. This chapter establishes the primary theoretical groundwork for the rest of the book. It first discusses the fundamental concepts of Chatman and Habermas, together with related concepts drawn from other theorists and researchers. Following this initial discussion, the chapter presents an overview of information worlds, drawing upon—and sometimes revising—Chatman and Habermas in order to build a coherent multi-level theory about the ways in which information intersects with and is perceived and used in many different social contexts, including not only formal institutions that are dedicated to making information available such as libraries but also more amorphous worlds in which the exchange and use of information is embedded in day-to-day life and social interaction.

ELFREDA CHATMAN, SMALL WORLDS, AND THE THEORY OF NORMATIVE BEHAVIOR

LIS theorist Elfreda Chatman worked throughout her career to develop a series of theoretical positions related to the place of information in the context of the everyday lives of definable localized social groupings of people, which she termed "small worlds," a phrase rooted in work describing phenomena ranging from large-scale social structures (Schutz & Luckmann, 1973; Travers & Milgram, 1969) to decision-making (Kilworth & Bernard, 1979) to various kinds of social networks (Kochen, 1989; Watts, 1999; Wilson, 1983; Wilson, 1999). Much of Chatman's early work focused on worlds specifically in settings constrained by socio-economic poverty as well as limited access to formal information resources, such as the small worlds of university custodians (Chatman, 1987), retired women in a residential retirement setting (Chatman, 1991a, 1992), and inmates of a women's prison (Chatman, 1999). In much of this early work, Chatman's concept of small worlds centered on the very sense of constraint and limitation found within them. Indeed, the very term "small world" seems to imply such a situation. In this sense, a world may be seen as "small" because of its perceived information poverty—life in a small world is one lived within the constraints imposed by either tight limits placed on the availability of information resources or a set of social norms that tends to discourage the world's inhabitants from looking beyond the boundaries of that world for the information they need.

However, the concept of the small world is, ultimately, neither evaluative nor absolutely yoked to poverty, whether socio-economic or informational. As a concept it is neither negative nor positive, but rather descriptive, acknowledging the "small" field of concerns and interests that are active in specific social settings and the predictability and routines of day-to-day life within those settings (Thompson, 2006, 2008). Late in her life, Chatman, with colleagues, turned her attention to small worlds which were far removed from the constrained and information-poor environments she had previously considered, applying the concept to settings such as virtual communities in information-rich online environments and feminist booksellers (Burnett, Besant, & Chatman, 2001). Whether the setting is marked by extreme limits on access to information or by the availability of extensive resources, members of a given small world perceive information in accordance with the norms of that world, and their behavior tends to be normative, "behavior which [they view] as most appropriate for [their] particular context" (Chatman, 2000, p. 13). A small world, whether it is geographically and economically constrained or enjoys access to a wealth of information resources, is *small* in the sense that its day-to-day activities and interests are structured and defined by a recognizable set of social norms and behaviors that are specific to the localized context of the world itself.

Small worlds, then, can be defined as the social environments in which an interconnected group of individuals live and work, bonded together by common interests, expectations, and behaviors, and often by economic status and geographic (or "virtual") proximity as well (Burnett, Besant, & Chatman, 2001). In these worlds, individuals may share similar opinions and concerns; similarly, the interests and activities of individuals are deeply influenced by the normative pressures of the small world as a whole (Chatman, 1999). Within each small world, everyday activities, including activities related to information, are thus considered to be "the way things are" and are frequently taken for granted as being standard across all small worlds, even when they are unique to a specific group.

Chatman's conceptualization of the small world is useful in that it explicitly accounts for the different ways in which people engage with and behave in relation to information within the context of their social interactions. While she presented several versions of the concept, the fullest account is to be found in her theory of normative behavior, presented late in her life (Chatman, 2000; Burnett, Besant, & Chatman, 2001). This theory is composed of four basic concepts: social norms, social types, worldview, and information behavior.

Social norms, the first of these concepts, refers to a world's shared sense of the appropriateness—the *rightness* or *wrongness*—of social appearances and observable behaviors. Social norms provide the members of a small world with a common understanding of the propriety of the visible social aspects of their world, influencing matters related to how individuals present themselves socially, from style of dress and conversational practices to work activities and patterns of consumption. They also define the degree to which it is or is not acceptable for an individual to step outside of the boundaries of their world and to interact with representatives of the outside world. Norms governing acceptable—or unacceptable—activities and behaviors may or may not cross boundaries between different small worlds. For instance, even a simple activity such as boisterous talking and laughing may be perfectly acceptable within one world while it can be something that raises suspicion and concern in another.

The second concept, **social types**, has to do with the ways in which individuals are perceived and socially defined within the context of their small world. An individual's role in a specific small world, thus, is in part a function of the ways in which they are "typed" or defined by other members of that world. Social typing takes place not only within the boundaries of a small world—that is, it controls how the inhabitants of a world understand each other—but also at points of intersection between a world and the larger society of which it is a part. Within a world, a specific individual may be seen by others as a trusted and reliable source of information, while information coming from another, because he is perceived as a troublemaker, will tend to be perceived as a disruptive influence. These perceptions may be entirely unrelated to the actual quality or reliability of the information itself. However, because the entrance of new information into a small world relies on outside sources (or, in the terminology of social network analysis, on "weak ties" between insiders and outsiders) (Granovetter, 1973), the social typecasting of outsiders has important implications for organizations such as libraries. For instance, a librarian who is tasked with providing information services to a community may be perceived as an untrustworthy outsider by the very community he or she is attempting to serve. Such outside information providers, especially if they disregard the social dynamics of the small world in which they are working, may face significant difficulties in making information available to that world.

Chatman defines her concept of **worldview** as "a collective perception held in common by members of a social world regarding those things that are deemed important or trivial" (Burnett, Besant, & Chatman, 2001, p. 537; see also Chatman, 1999). Members of a small world are not only united by a shared set of normative behaviors (as the concept of social norms describes) but also by a common understanding of which aspects of the world—both their own small world and the greater world beyond its boundaries—are of value and are worth their attention. Worldview, thus, relates to the "scope" of a small world, establishing the relative importance of beliefs and bits of information that are available to members of the world. Thus, while social norms have to do with the *propriety* of observable manifestations of a world's beliefs, worldview influences the degree to which a world's members perceive those beliefs to be important or trivial. Worldview, similarly, influences whether or not people are interested in events and issues outside of the boundaries of their own small worlds, as well as their understanding of which outside events are worth their attention. Thus, while two groups may share an interest in international news, one may focus on Cuba, while the other may focus on Israel, each group largely disregarding news of interest to the other.

The final concept of the theory of normative behavior, **information behavior**, concerns the uses to which information is—or is not—put within a small world and refers to the full range of possible normative behaviors (as regards information) that are available to its members. Historically, LIS research has focused almost exclusively on only one type of information behavior, "information seeking behavior," in which an individual, interacting with a formal information service such as a library, presents an explicit information need formalized as a query (Case, 2002). By comparison, the concept of information behavior acknowledges that people interact with information in a wide variety of ways, from the informal exchange of information among friends, to online browsing, to posting fliers in grocery stores, and even to the active avoidance of information that is somehow deemed to be inappropriate or dangerous. Such activities are not necessarily tied to "objective" measures of the usefulness or value of the information itself. For instance, in a specific small world, information about a particular style of music may be enthusiastically sought and exchanged, while certain health information may be dismissed or censored because it is considered to be socially unacceptable.

Chatman's theoretical work is useful for the ways in which it conceptualizes the place of information within specific localized small worlds. Chatman's approach fits well with other disciplines' findings regarding the values placed on information. Studies in sociology and psychology indicate that individuals experience culture as fragments of information but that culture also serves to give structure to the fragments of information (DiMaggio, 1997; Martin, 1992). Among these fragments, people are more likely to recall information correctly and efficiently if it fits within their established cultural frameworks (DiMaggio, 1997). Chatman's findings about the uses of information in social contexts are also reflected in studies in other fields. A series of studies of government agricultural agents in rural Africa, for example, found that agents who presented the information they were distributing in a manner that accounted for the highly localized beliefs of the farmers were far more successful than other agents at getting new agricultural practices adopted and implemented (Woods, 1993).

However, because it draws such tight circles around small worlds and rarely looks at information beyond those boundaries, Chatman's work does not adequately consider interactions between small worlds and the broader society within which they exist, nor does it account for interactions across and between multiple small worlds. Not only do individuals, as they move between different parts of their lives—from neighborhood to work to shopping, for instance—encounter many different norms and behaviors, but they are also subject, in many cases, to forces and influences from outside of the specific localized world they inhabit at any one time. As Truman (1971, p. 509) has noted, "no tolerably normal person is totally absorbed in any group in which he participates. The diversity of an individual's experiences and his attendant interests involve him in a variety of actual and potential groups." By contrast, the work of Jürgen Habermas, and particularly his concept of the lifeworld, focuses almost exclusively on the social world and its information resources in a much broader context.

HABERMAS AND THE LIFEWORLD

While the influence of Elfreda Chatman's ideas has largely been limited to the world of LIS, Habermas has been widely influential across many disciplines, spawning a considerable body of secondary literature (i.e., Alejandro, 1993; Clark, 2000; Zaret, 2000, among many others). Scholars in fields as diverse as political science, communication, public policy, cultural studies, and education have explored the ramifications of his ideas for their disciplines, including the concepts of the public sphere, lifeworlds, and ideal speech situations.

In LIS, however, Habermas' work has not been widely examined; those writers who have considered his ideas have, for the most part, used them in discussions of users of libraries, of staff and managers of libraries, or of the operational context of libraries (Buschman, 2003; Wiegand, 2005). Although his ideas have important implications for understanding the place of information within culture and society, their significance for issues related to information behavior remain largely unexplored.

Central to Habermas' work is the concept of the public sphere, an idealized "space within a society," essential to the functioning of a democracy, which is "independent both of state power and of corporate influence, within which information can freely flow and debate on matters of public, civic concern can openly proceed" (Corner, 1995, p. 42). Habermas' writings detail the exchange of information in the larger social and political processes of a society in relation to public institutions and forums, such as public libraries and independent news sources (Habermas, 1989). Habermas believed that democracy was not possible without public participation and critique, and this participation has to occur in public forums to be truly effective. The public sphere "may be conceived above all else as the sphere of private people come together as a public" (Habermas, 1989, p. 27).

Habermas suggested that the public sphere first emerged in eighteenth-century England where, in coffee houses and salons, a mix of citizens from the mercantile classes and the public press and others could, regardless of social status, gather to publicly discuss, analyze, and criticize government actions and policies. In Habermas' scheme, topics of concern within the public sphere are those of social and political consequence for society as a whole as well as for individuals. An active public sphere, thus, functions as an important link between the members of a democratic society and the government.

Habermas conceived of the "authority of opinion" in the public sphere as a "precondition" for true liberal democracy (Zaret, 2000, pp. 21–22). The locales and communication channels of the public sphere—the public press, forums, schools, libraries, and other settings—not only make free discourse about social and political information possible, they also function as mediators between the rights of the individual and the power of the state in democratic societies. The public sphere became clearly differentiated once political opinion could be found in mass, rather than just elite, communication (Giddens, 1985). As the public sphere grew in importance, legal rights were established to protect it and to secure its role in democratic participation (Habermas, 1989). Over time, the public sphere became essential "to the protection of the civil liberties that are considered essential in modern democracy" (Nerone, 1994, p. 6). The guarantees of freedom of speech, expression, press, and assembly in the Constitution and Bill of Rights of the United States reflect the importance of protections for open discourse about meaningful social and political issues. The free flow of information also reflects how the founders designed the Constitution to encourage citizens to be involved in their communities and become true citizens by forcing involvement to check others' ambitions and advocate for their own perspectives (Goldwin, 1986).

For the public sphere to successfully support the exchange of information necessary for a healthy democracy, it must feature open communication, information access, and information exchange (Burnett & Jaeger, 2008). The role of ICTs in democracy has become increasingly vital as the size of populations has grown. "Pure democracy" in the sense of the Greeks or Rousseau is impossible for a modern society—200 million people "would *never* get through talking" (Schattschneider, 1969, p. 61). In such a sphere, people are able to interact freely with one another; to gain access to authoritative and reliable information resources; and to openly exchange information between and among each other, independent of formal and official channels of communication and information distribution such as the mass media and governmental information services (Murdock & Golding, 1989).

A strongly functioning public sphere can, at least in the abstract, protect the communication practices and the open exchange of information from both political and corporate influences. However, the role of the public sphere has changed—and even been eroded—over time. Limitations to the power of the public sphere have emerged from governmental actions designed to limit the access and exchange of social and political information in some social settings (Jaeger, 2007; Jaeger & Burnett, 2005), as well as from rapid increases in mass media control of communication channels (Hiebert, 2005; Nerone, 1994; Starr, 2004). In other instances, government and mass media have joined forces to impose limitations on the public sphere (Ewen, 1996; Hiebert, 2003, 2005). The impact of such limitations on the information behavior of social groups and individuals remains largely unexplored. However, Habermas believed that governmental and corporate controls were eroding the public sphere, with members of society becoming acclimated to serving as politically unaware followers of government and corporations, ultimately leading individuals to focus on small personal interests rather than large social and political ones (Habermas, 1992).

Closely related to the public sphere—and key for the discussion at hand—is Habermas' concept of the information lifeworld, which can be defined as "the whole ensemble of human relations which is coordinated and reproduced" through communication practices and information exchange (Brand, 1990, p. xii). A lifeworld "stands behind the back of each participant in communication" and "provides resources for the resolution of problems of understanding" (Habermas, 1992, pp. 108–109). Further, unlike the strictly localized small scale of Chatman's concept of the small world, Habermas' lifeworld is expansive, reaching across a broad swath of a culture: "members of a social collective normally share a lifeworld" (Habermas, 1992, p. 109). Others have used the term "lifeworld" in alternate ways, such as Edmund Husserl (1962, 1970), who uses it to describe the phenomenological background of an individual's experience. However, the discussion in this book will focus on the lifeworld in the Habermasian sense.

A lifeworld, then, is that collective information and social environment that weaves together the diverse information resources, voices, and perspectives of all of the members of a society. In this increasingly technologically mediated information environment, the lifeworld can be seen as the totality of communication and information options and outlets available culturewide. It is a dizzying array, a mass of traditional and new media, channels, and services, comprising television and radio, news and entertainment, blogs and supermarket bulletin boards, virtual communities, and much more. It encompasses the full range of such resources. The concept of the lifeworld does not focus, unlike Chatman's concept of the small world, on the specifics and contextual aspects of localized communities. To Chatman's necessary little picture, it provides the equally necessary big picture.

Within the lifeworld and the public sphere, Habermas conceived further of an idealized, theoretically optimal form of communication that, in the abstract, would be the most productive and the most useful for fulfilling the promise of the public sphere: the ideal speech situation. This can be defined, as Alejandro (1993) put it, as "a conversation among equal individuals with each participant having equal opportunities to assert and dispute" (p. 187). As an idealization, the ideal speech situation supports discourse rooted in well reasoned and effectively argued critical analysis, leading to consensus within the public sphere and across the lifeworld. Such discourse focuses not on the individual speaker in context, but on the effective transmission of accurate information across contexts. "In the public sphere, reason, not passion, and not personality, must govern" (Nerone, 1994, p. 5). With a commitment to a democratic society and the ideals of truth, free access to information, and justice, "ideal speech is inconsistent with an intention to distort, or use overweening power or wealth purposely to manipulate" (Price, 1995, p. 25).

Further, for ideal speech situations to be realized, the public sphere and the lifeworld must be sheltered from intrusions into the workings of communication and information channels and from political and corporate efforts to establish control over the content of speech and to place limits on access to information (Zaret, 2000). Corporate control of media channels and information resources—and governmental intrusion into public sphere activities—should be viewed as the "colonization" of the lifeworld, inexorably leading to "the impoverishment of expressive and communicative possibilities" (Habermas, 1984, p. 20). Such colonization, whether originating in the profit motives of the corporate marketplace or in the political manipulations of a government, reduces access to and the exchange of political and social information, placing serious limits on the numbers of voices which can be heard, and thus further constraining the effectiveness of the public sphere. In Habermas' scheme, once such a set of limits and constraints are in place—often through overt, deliberate governmental actions—the eroded public sphere is nudged into a position of "remoteness from the political system," severely undermining the ability of members of the society to have any impact on either the workings of the marketplace or on the functioning of their governments. Although Habermas' focus was on the society as a whole, such governmental and corporate activities can also have a significant impact on information and communication within small worlds. Because their focus is near to home, centering on their localized context, inhabitants of specific small worlds may be unaware of the constraints imposed on them from outside.

The concept of the public sphere has been widely analyzed, applied, and criticized. For the purposes at hand, several of the critiques are of particular interest. First, Habermas viewed the public sphere as "the world of communal, economic, and political life rather than intimacy and familial relations" (Green, 2001, p. 117). However, it has been noted that "it is possible to have private interests, and still have a public one" (Alejandro, 1993, p. 184). Habermas may have too rigidly conceived of the public sphere being separated from personal interactions. "The only requirement in Habermas' model is that participants are able to justify and redeem their objections via the force of the better argument" (Clark, 2000, p. 49). As such, the interrelationships between the small worlds of a personal level and the political concerns of the lifeworld expressed through the public sphere are by no means incompatible.

Second, the continuing viability of the public sphere has received a great deal of attention. Many believe that societies that once supported a more vibrant public sphere "have experienced a decline of meaningful public discourse, [but] they nonetheless contain the potential for the reconstruction of critical and rational forms of public communication" (Clark, 2000, p. 43). A key concern has been that reductions in the number of venues of the public sphere and the government and media controls over the public sphere have reduced the ability of citizens to participate as they should. "Political equality without deliberation is not of much use, for it amounts to nothing more than power without the opportunity to think about how that power ought to be exercised" (Fishkin, 1991, p. 36). It has even been suggested that modern democracy is no longer an exchange of ideas to formulate the will of the people, but a series of options presented to consumers (Dahlgren, 1995). Habermas, however, viewed modernity as an unfinished project that could still aspire to ideals of early democratic discourse (Habermas, 1996b). Even if the public sphere is not as robust as it may have been in the past, it may increase in viability with improved access and exchange of information.

Ultimately, Habermas believed the public sphere aspired to be a conversation among equals rather than a hegemonic tool. The public sphere is "a conversation among equal individuals with each participant having equal opportunities to assert and dispute" (Alejandro, 1993, p. 187). The equality of voices is essential to the discussion at hand. Though Foucault (1979) believed that reason could not function as a tool of oppression, many recent events have demonstrated that discourse can be distorted by hegemonic influences. However, the strength of the public sphere "lies in the presumption of reason, the human ability to define and solve problems" (Boeder, 2005, n.p.). The negative turn of political actors successfully stifling the public sphere represents a serious and pressing challenge to the public sphere, but it does not undermine the potential power it can have in deliberations about social and political issues. While hegemonic elements of societies tend to try to enforce a dominant or common culture, every society is in fact comprised of many different cultures, just as the lifeworld is comprised of many different small worlds (Williams, 1958, 1968). So long as societies encompass a range of cultures—a range of small worlds—the public sphere will be a viable concept.

BEYOND CHATMAN AND HABERMAS: INFORMATION WORLDS

The concepts of both Chatman and Habermas are useful tools for analyzing the social contexts of information and information use. However, each is problematic when taken in isolation. Chatman's work does not explicitly consider either worlds beyond the boundaries of localized contexts or the relationships between multiple worlds. The unique small worlds are vital to understanding the normative behaviors and choices of individuals who live within them; however, it also must be recognized that small worlds are situated within a larger lifeworld, and that there are boundaries between worlds across which individuals may move. The worlds that an individual lives in do not all share the same normative behaviors. People raised in these different communities will likely engage in different normative behaviors and hold different socio-cultural beliefs and values. This may in turn impact their decisions as they probe the boundaries of their own worlds, explore other small worlds, and cross into the larger lifeworld. Similarly, Habermas' conceptualization of the lifeworld, which is extremely useful for considering the ways in which information plays a role across the political and social landscape of a culture at large, not only disregards localized contexts—Chatman's small worlds—but also focuses nearly exclusively on information related to politics and social issues broadly conceived, actively discounting information related to more quotidian aspects of peoples' lives.

However, a more general theory combining concepts drawn from each—the theory of information worlds detailed in this book—can provide a more widely applicable and robust model for examining the ways in which information functions within the social worlds of people. Such a combination, further, can provide some useful correctives for some of the limitations inherent in both Chatman and Habermas.

In the broadest sense, the need for a theory of information worlds grows out of the fact that neither Chatman nor Habermas provide conceptualizations of social worlds beyond their relatively narrow foci. As was noted earlier, Chatman limits her consideration to very specific localized communities, while Habermas' lifeworld concept encompasses the sum total of information and communications resources across a society as a whole.

Each of the two largely disregards issues they take to be outside of their scope. Chatman rarely considers whatever other worlds are to be found outside of a specific small world, whether the broader social context within which a small world exists or other small worlds, even when those multiple worlds come into contact with one another; Habermas, conversely, does not investigate how the broader lifeworld might be instantiated within or might interact with localized contexts and specific communities (Burnett & Jaeger, 2008). The theory of information worlds, by comparison, melds these two approaches into a single multi-leveled conceptualization of the interactions between social norms and values, information, and community, particularly in situations in which multiple small worlds overlap.

The theory of information worlds explicitly acknowledges that social contexts are not, in most cases, isolated from one another. To take only the most obvious example, each of Chatman's small worlds, however constrained or bounded it may be by economic or information poverty (or by other restraints, whether geographic, social, or legal), still exists within a broader social context. No matter how heavily drawn, the conceptual circle that encloses the small world, separating it from the world around, is equally a boundary and an interface. While, as Chatman argues, the day-to-day activities and attitudes of a small world tend to be normative and specific to that world, they are not utterly unique or separable from the world at large but are, willy-nilly, impacted by a wide variety of influences, ranging from global, regional, and local economic and political forces, to the media and the range of available technologies, and even to factors such as details of the physical environment and patterns of transportation and urban design.

Conversely, while Habermas' lifeworld pervades a society, functioning as a kind of cumulative store of information and communication resources, the theory of information worlds explicitly acknowledges that a lifeworld is not the same in different contexts and in different locales within the society. Nor are the concerns of the public sphere and the broad lifeworld, linked as they are to social issues and events of wide significance, always the concerns of significance or importance within individual worlds, which may be focused on issues that appear trivial or mundane when viewed through the lens of the lifeworld. That is, even though the lifeworld is, theoretically, shared by all small worlds, it is perceived differently within different small worlds and is instantiated or realized differently, depending on the localized context within which it is experienced by individuals and communities. Further, individual worlds may even invest matters that seem trivial in comparison to more pressing social concerns—the endlessly unfolding plots of television soap operas or the staged competition of a television show such as *American Idol*, for instance—with attention far outstripping what is considered appropriate in other worlds. One community's experience of the shared lifeworld is, to put it bluntly, often incommensurate with that of a different small world.

In this sense, information coming from the lifeworld can be conceived of as a kind of "boundary object" (Bowker & Star, 1999; Star & Griesemer, 1989). It is something that crosses "several communities of practice" and is both "plastic enough to adapt to local needs and constraints" and "robust enough to maintain a common identity across sites" (Bowker & Star, 1999, p. 297). As a boundary object, lifeworld information exists both apart from individual small worlds and also within individual worlds, where it is instantiated and experienced according to the constraints, expectations, and norms of the localized worlds themselves. Similarly, agencies of the public sphere are not the only channels through which significant information can be accessed and shared. In addition to such public sphere channels and the perhaps compromised channels of traditional information sources such as the media or the formal information (or propaganda) efforts of governmental and other agencies, people receive information through interpersonal interactions that may otherwise seem trivial or quotidian, whether through chitchat or through conversations that seem to occur as a side effect of other activities. For instance, LIS researchers have identified many "information grounds"—locations where people gather for purposes unrelated to information, such as a barbershop or a medical clinic, yet nevertheless engage in networking to facilitate formal and informal information exchange (Fisher, Durance, & Hinton, 2004; Fisher & Naumer, 2005; Fisher, Naumer, Durrance, Stromski, & Christiansen, 2005).

While it draws its conception of small worlds from Chatman and its conception of the broader lifeworld from Habermas, the theory of information worlds also acknowledges that there are intermediate worlds, which can mediate or intervene between the macro and the micro. For instance, in the contexts of formal institutions and corporate life, researchers in the field of "new institutionalism" argue that there is a "meso" or organizational level that establishes and enforces norms within institutional contexts, and that also translates or contextualizes macro-level norms and information resources for use within the small worlds of an institution (DiMaggio, 1997; Powell & DiMaggio, 1991). Just as small worlds and the lifeworld can be conceptualized independently, "organizations have an autonomy both from the societal structures in which they are located and from the individuals who compose them" (Friedland & Alford, 1991, p. 241). However, the theory of information worlds suggests that there are interactions between all of these levels—small worlds exist within a broader lifeworld context, which influences them, just as the lifeworld itself can be influenced by specific small worlds, and intermediate levels can interact with both.

In this sense, the theory of information worlds also reflects some of the same goals of the theory of structuration, though the ultimate purposes are quite different. The theory of structuration was intended to provide "a conceptual scheme that allows one to understand both how actors are at the same time the creators of social systems yet created by them" and a "conceptual means of analyzing the often delicate and subtle interlacings of reflexively organized action and institutional constraint" (Giddens, 1991, p. 204). As such, the intent of the theory of information worlds to demonstrate and analyze the interrelationships and simultaneous interactions between the small worlds of information and the information in the public sphere parallels the attempt of the theory of structuration to explore the dual roles of individuals in society.

The theory of information worlds provides a framework for understanding the multiple interactions between information and the many different social contexts within which it exists, from the macro (the lifeworld) to the meso (intermediate) to the micro (small worlds). It asserts that information behavior is shaped simultaneously by both immediate influences, such as friends, family, co-workers, and trusted information sources of the small worlds in which the individual lives, as well as larger social influences, including public sphere institutions, media, technology, and politics. These levels, though separate, do not function in isolation, and to ignore any level in examining information behavior results in an incomplete picture of the social contexts of the information.

CORE CONCEPTS OF INFORMATION WORLDS

In terms of core concepts, the theory of information worlds borrows the most heavily from Chatman's work. Of her four fundamental concepts, the theory of information worlds integrates and extends three as-is: social norms, social types, and information behavior. The fourth, worldview, is presented in somewhat modified form and re-named **information value**. Further, because this theory focuses not only on discrete isolated worlds but also on the interactions between and among worlds, it requires an additional concept: different worlds are separated by **boundaries**, through which communication and information exchange can—but may or may not—take place. From Habermas, the primary structural borrowing is the concept of the lifeworld, which serves as the overarching world within which small and intermediate worlds exist, providing a kind of meta-context for analyzing the place and activities of multiple small worlds. Further, the concept of the public sphere has significance for this theory, particularly for ensuing discussions of political issues and other issues of broad social concern.

Some further definition and discussion of these concepts is necessary at this point, as is some consideration of the term "information worlds." The use of this term in this book follows a great deal of thought. A main goal was identifying a name that would not be problematic in relation to other widely established terms in social theory across disciplines, such as the problem with the numerous different established uses of "small worlds." As with "small worlds," "information worlds" (or "information world") has been used in a range of different scholarly fields to mean several different things. Curiously, these previous uses have typically not included clear definitions of the term. All of the previous uses of the term, however, have relevance to the meaning of information worlds within the context of the theory detailed in this book.

The term has been used in a number of fields as a causal term to describe the information in a particular social, physical, or virtual environment or place. Liesener (1983), for example, uses "information world" to describe the technological richness of schools in the 1980s. Streitz, Magerkurth, Prante, & Rocker (2005), from the perspective of design, approach information worlds as the physical or virtual environments in which a user could gather information. Others have used the term to describe electronic parts of management structures in business literature (Rayport & Sviokla, 1995; Walsh, 1988). Heiskanen (2002) conceives of information worlds as protected, isolated spheres used to keep certain information secret, such as proprietary information of a corporation.

More common have been usages of the term to describe various online environments, such as Roesler and Hawkins (1994), who refer to the Internet as an electronic information world. Kunii (2000) uses the term synonymously with cyberworlds, which he describes as the creation of hypothetical materials in computers. Jul and Furnas (1998) uses information worlds interchangeably with electronic worlds to describe interactive online environments. To Rennison (1994), they are parts of large, abstract datasets that could be used in the visualization of electronic information.

The closest usage of the term "information worlds" to the way that it is used in this book derives from the use of the term by Elfreda Chatman herself, who (1987, 1992) twice used "information world" in the title of a publication, each of which focused on a very specific social group—low-skilled workers and retired women. The term is also used in the text of two other of her works (Chatman, 1991b; Chatman & Pendleton, 1995). In these cases, Chatman does not make an attempt to define or operationalize the term. From her use of the term, however, she appears to have intended it to signify the sources of information—mostly social—that individuals encounter and engage in their everyday life activities, with a particular emphasis on information gathering.

As with other concepts Chatman used, her focus was on small social units when discussing an information world, and a number of scholars have followed Chatman's lead in the use of the term to describe the information available to specific, small social units. A line of research about digital libraries and communities of practice, for example, discussed the information worlds associated with communities of practice (Bishop, 1999; Bishop, Neumann, Star, Merkel, Ignacio, & Sandusky, 2000; Star, Bowker, & Bishop, 2003). Similarly, Lu (2007) followed Chatman's use of the term to frame a discussion of the process of information acquisition.

While these various uses of "information world" and "information worlds" to indicate different concepts are a sampling, they represent the range of different uses in many scholarly discourses. Intriguingly, all of these conceptualizations of information worlds fit into the broad concept of information worlds at the heart of the theory detailed in this book. The theory of information worlds embraces most of the social, technical, physical, and virtual dimensions that are variously discussed in these works. Application of the theory of information worlds would be appropriate in most or all of the research contexts detailed here, emphasizing a key asset of the theory of information worlds in addressing problems of information.

The concept of the lifeworld—like the three concepts borrowed from Chatman—is used without significant modification. Situated within the broader lifeworld, multiple small worlds (as well as intermediate worlds such as the meso-level worlds of the new institutionalism) exist, and individuals carry out their day-to-day activities within the boundaries of their small worlds (and individuals will most likely inhabit multiple small worlds, though their activities at any given time probably occur within the bounds of a single world). Within a given world, as Chatman notes, visible behaviors are governed by and understood according to the specifications of **social norms**; actions are undertaken by specific social actors, who embody the **social types** used to define them within their world; and the patterns of information exchange and use are regulated by the world's shared understanding of appropriate **information behavior**. Following Chatman, behaviors reflecting these three concepts can be found within the context of any information world.

Chatman's concept of **worldview**, however, is problematic, in large part because the term has been widely used in other contexts for often radically different purposes. There are two key problems with worldview in this context. First, it has been used in many different disciplines, ranging from philosophy to linguistics to literature to cognitive science, without a great deal of consensus as to meaning or implications. Historians of Stonehenge, for example, believe that the architecture of the monument reflects a worldview of the relationship between the earth and the sky that was ultimately expressed in a sixteen-month calendar (Davies, 1999; Thom, 1967). The range of uses—and inherent imprecision of the term—can be seen through the website http://www.projectworldview.org. Second, and more significantly, worldview typically is meant to indicate a perspective developed over an extended period of time, often tied to the long-term development of language, geography, economics, and resources across generations (Naugle, 2002). Worldview has even come to be seen as including the perspectives of different religious groups (Kitaro, 1987; Naugle, 2002). The use Chatman intended was much more limited, focusing on the shared understandings of small groups of people living and working together in definable and bounded cultural worlds. As such, Chatman's use of the term does not fit well with the general use of this concept in other disciplines, and the disagreements about and wide range of uses of the concept undermine its utility in this context.

In place of worldview, this theory employs a concept of **information value**. At its root, this term is defined much as Chatman defined worldview: within a world, it designates a shared sense of a relative scale of the importance of information, of whether particular kinds of information are worth one's attention or not. However, the term has been chosen, in part, because it also reflects the different kinds of value that different worlds may attach to information: information may, for instance, have emotional, spiritual, cultural, political, or economic value—or some combination—within a specific world. In the broader lifeworld, information is always embedded within some kind of economic and political system and is always charged with political and economic significance that may be quite different from the sense of worldview designated by Chatman's concept. Clashes between different information worlds can often result from differences in perceived information value between worlds, well beyond simple differences in the perceived importance of information—when information is of political and/or economic value, such clashes can come to be of global importance. This concept will be explored at length in the next chapter.

As noted in passing earlier, the term "small worlds" has itself been used with varying definitions in other contexts as well. These range from the notion of "six degrees of separation," a concept drawn from social network studies that has entered into the popular vernacular (for just one example of an early study of such small worlds among many, see Milgram, 1967; for an account of how the concept has entered popular culture as "Six Degrees of Kevin Bacon," see Wikipedia, 2008). Further, the term has been used in formal studies of many different kinds of "small world networks," or coupled, dynamic systems that include everything from power grids to neural networks of worms, from the spread of disease to collaboration graphs of actors that are derived from mathematical analysis (Kretschmar & Morris, 1996; Watts, 1999; Watts & Strogatz, 1998). In a social context, such a concept "allows better prediction and understanding of message flows, efficiency in information search and retrieval, and the degree of social stability among members of the network" (Ravid & Rafaeli, 2004, n.p.).

In this version of the small world concept, a network is seen as a system filled with hubs with many connections and cliques that develop where a group of connections is very strong. In terms of members of a social network, each node would be a member, the edges would connect members that know one another, members that know many other members would be hubs, and members that do not know each other would be connected through other members who know them. As a result, searching for information or connecting with others can be more efficient in small world networks, though the distance between members grows in proportion to the number of members in the network (Comellas & Sampels, 2002; Ravid & Rafaeli, 2004). Interestingly (particularly for the discussion of technologies in Chapter 5), the technologies that are essential for virtual communities and other online social networks are strict small world networks, connecting participants through sets of nodes, hubs, and other physical connections (Buchanan, 2003; Dorogovetsev & Mendes, 2003).

Although it substitutes the term "information value" for "worldview" because it, like "small worlds" has been widely used in other senses, the theory of information worlds retains small worlds as a core component for two reasons. First, unlike worldview, other uses of small worlds parallel Chatman's use of the term; that is, they all point to definable—and, in some cases, measurable—social sets of participants, emphasizing the interactions between those participants. In other words, they are all linked to *networks*, and to the flow of information and interaction within those networks and between different networks. Second, the term works well for the kinds of social environments explored by Chatman; that is, there are information worlds of many different sizes, ranging from the small, localized worlds within which people conduct their day-to-day activities and live their lives, there are larger social worlds—the meso level of new institutionalism—that may contain several distinct smaller worlds, and there is the large, overarching information world of the lifeworld. Within the broader context of a theory of information worlds, Chatman's term retains an appropriate and useful acknowledgment of relative scale.

Chatman's work, in practice, focused only on individual small worlds; she did acknowledge that multiple such worlds existed simultaneously, although she actively downplayed the degree to which interaction between different worlds was possible. In her work, within any given small world, there are basically only two positions: one can be inside of that world or one can be outside of it. However, as in Habermas' notion of the lifeworld, these multiple worlds not only function as separable entities unto themselves, but also form part of a larger whole. Further, different small worlds, as parts of that larger whole, do come into constant contact with each other in many different ways. Not only can an individual be, at different times, a member of multiple worlds, but he or she can, under some conditions, act as intermediary—or gatekeeper—between those worlds. The individual can carry information from one to the other, or can protect one world from intrusions by the other, because of an understanding of the two worlds' differing social norms and expectations.

In addition, because they co-exist within a larger social and informational context, different worlds can come into contact with each other because of overlaps in the spaces they inhabit or because they share certain interests and sets of social norms. The can also—and often do—break into open conflict with one another because of *differences* in those interests and social norms or because they have different understandings of information that they have a mutual interest in. For instance, Burnett, Jaeger, and Thompson (2008) investigated three different instances in which different small worlds clashed over the different values and understandings with which they invested shared resources or social settings. Because of such interaction and conflict, the theory of information worlds makes use of a concept to describe the points of contact between worlds: information worlds are separated one from another by boundaries. Further, small worlds with the greatest levels of access to information do not necessarily have the highest levels of influence when they come into contact with one another (Kim, Lee, & Menon, 2008).

Information worlds may, like cities or states, be wholly enclosed by other, larger worlds, but they may still maintain their own identity through a definable boundary that surrounds them and marks their identity apart from the larger world of which they are a part. They may also, like nations, share boundaries—either long or short—that both separate them and keep them in contact. Like political borders, the boundaries between different information worlds can take a number of forms. They may be permeable or carefully guarded, they may be sites of conflict or cooperation, they may be mutually agreed-upon or contested, or they may designate natural differences between worlds or be artificial, among innumerable other variables.

A METAPHOR FOR INFORMATION WORLDS

It has been rightly noted that it is impossible to "draw exact blueprints for systems whose components are people and institutions in pursuit of their own goals and interests" (Swanson, 1980, p. 84). Given that the theory at hand is an attempt to create a framework for such a large-scale problem, to fully understand the theory and its constituent aspects and relationships it may be helpful to visualize a representation of what information worlds might look like.

Imagine a sink full of soap bubbles. Each individual bubble is a small world with its own social norms, social types, information value, and information behavior. Within each bubble, the members of that small world have established ways in which information is accessed, understood, and exchanged. However, each bubble also touches many other bubbles at its edges. These boundaries between the soap bubbles represent points of contact between different small worlds. Few individuals exist only in one small world; a typical person is a part of many small worlds—friends, family, co-workers, people with shared hobbies, etc. Where a bubble touches another exists a boundary between two small worlds. And new bubbles are constantly being created as the soap is mixed.

Information moves through these boundaries via people who are members of these two worlds, through channels of communication, or through interaction between members of two small worlds in a place where members of different small worlds are exposed to other perspectives, such as in a public library. As information moves through boundaries between small worlds, the information is valued, treated, understood, and used differently in each small world in line with the social norms of that world. So, the same piece of information may have a different role within each bubble.

However, these bubbles collectively also constitute a sink full of bubbles. The world of the sink is thus comparable to the lifeworld. The way that the bubbles as a group treat information will shape how the information is treated across the sink as a whole. As the information moves between bubbles, more and more small worlds will decide how to treat this information, creating an overall perception of the information across the lifeworld. Groups of bubbles building a collectively shared value for the information create meso-level information worlds.

There are also influences that interject into the lifeworld to shape how the small worlds treat information. Public sphere organizations—like schools and libraries—exist specifically to ensure that information continues to move between bubbles and that members of each small world are exposed to other small worlds. In this sense, public sphere organizations act like sides of the sink, keeping the bubbles from floating out of contact with one another. In contrast, certain influential small worlds—those who possess political power or those who control the media, for example—can use their power to push back against the collective small worlds to enforce a minority perception on the majority, much like turning on the faucet over the sink. Finally, like additional soap being added to the sink, ICTs act as a way for small worlds to connect in new ways and to reach other small worlds that would not otherwise touch their boundaries. The Internet and online social networks represent particularly powerful examples of this phenomenon. In total, the small worlds are shaped by all of these larger influences, but also have the power collectively to define the parameters of the external influences.

As such, the sink and all its bubbly contents represent an information world in the largest sense. In addition, many smaller and intermediate information worlds also exist within the sink as related clusters of bubbles that are tied together in some familial, community, professional, educational, social, cultural, political, geographical, technological, or other means.

This (admittedly somewhat whimsical) metaphor also distinguishes the theory of information worlds from its antecedents. Chatman looked exclusively at individual bubbles, disregarding everything else in the sink. Habermas, on the other hand, only was interested in the sink, not its contents. The theory of information worlds, however, attempts to account for all of the elements at work.

The remainder of this book—first by expanding the theory of information worlds and its concepts and then by applying it to specific contexts—will focus on the permutations of the theory in terms of the components of, influences on, and contexts for information worlds. While the traditional venues of the public sphere have mostly eroded—with the clear exception of the still vibrant public library—there are many contexts in which information worlds thrive. These contexts range from the new technological venues on the Internet such as blogs, wikis, and discussion forums to the activist political realm to the plentiful worlds where information exchange is embedded in day-to-day life and social interaction.

3 Information Value

As was detailed in the preceding chapters, the work of Elfreda Chatman is, along with that of Jürgen Habermas, at the heart of a robust conceptualization of information worlds. However, for a number of reasons, her use of the concept of worldview as one of the four foundational concepts of the theory of normative behavior proves to be problematic, in part because of the term's quite different usage in other contexts, and in part because of her narrow focus on isolated and independent small worlds. As a result, this theoretical approach to information worlds replaces her concept of worldview by positing a concept of information value, itself a complex notion intended to provide a framework within which to examine the multiple roles played by information—and the value or values associated with it—in varying social contexts. This chapter presents a full discussion of the concept of information value, delineating its various dimensions and applying it to several social settings.

In its simplest outlines, the concept of information value can be seen in terms of a framework that can be broken down into two component parts:

1. Content and
2. Control

This framework dominates much of the discussion of information in library and other settings focused on providing information to users. For instance, one classic model used to explain information seeking and information services relies heavily on defining two discrete stages of a single process. In this model, an information seeker approaches an information service with an information need that is ultimately, through a process of negotiation, expressed as a formal query and presented to an information system such as a catalog, a database, or a library collection. In response, the system provides a set of responses to the query that ideally, in one way or another, match the information need as expressed in the query. From beginning to end, the informational heart of this process is overwhelmingly a matter of information content—the patron needs information that is about a particular issue in his/her life, and the system, if all is successful, supplies information that matches that need in terms of content (Case, 2002). Indeed, in the field of LIS, the entire concept of the *relevance* of information is tightly interwoven with concepts of content and aboutness.

Such an explicit focus on the quality and pertinence of content per se can also be seen in other settings as well, often as a selling point. For instance, the famous motto of the *New York Times*, "All the news that's fit to print," can—and most likely should—be read as an explicit claim about content as information value. An accurate paraphrase might read "if there's information about an important issue, you'll find it here." The currency of information, of course, has an impact here as well, though not all information that is relevant at the current moment makes it into such a "paper of record." Only certain types of informational content are considered appropriate for inclusion, and readers expect to see only a relatively narrow range of topics and aboutness addressed in the pages of a paper like the *New York Times*.

The second term in this view of information value as a duality also has important resonance within the library tradition, although control has very different—and equally key—connotations in other settings. In the information world of the library, "control" most directly refers to those mechanisms and procedures put in place to ensure quality control and consistency in the bibliographic records representing individual items in collections. As a 2008 report to the Library of Congress put it, "Bibliographic control is the organization of library materials to facilitate discovery, management, identification, and access. It is as old as libraries themselves . . ." (Library of Congress Working Group, 2008, p. 6). Organizational tools and approaches for control of information include such well-known mechanisms and supports as authority control files for the presentation of authors' names, and rules for the treatment of titles. In the library world—and in many academic circles—such control has also been treated in terms of authority, often being put forward as a fundamental argument in favor of the use of libraries as holders of authoritative and reliable information resources against the more free-wheeling and uncontrolled environment of the Internet.

Such a definition is inescapably linked to quality control. In the library tradition, emphasizing as it does open access to information, control is almost entirely associated with quality and consistency—information is controlled expressly in order to guarantee its quality and to ensure that searches are both consistent and accurate. In other settings, however, control carries quite different implications, being associated not with processes and practices designed to ensure quality, but rather with overt or covert attempts to place restrictions on access to information. Perhaps the most notable context within which such a sense of control predominates is the political realm, where information can be perceived along a continuum in terms of who should be allowed access and who should not. As will be discussed at length in Chapter 8, control of access to information plays an extremely important role in the information world of politics, with access to particular bits of information either being controlled (that is, limited) to mitigate the perceived dangers of allowing access or being disseminated as a tool in a political struggle.

Questions about the control of information, however, are also critical in analyses of other information phenomena, including the media (discussed in Chapter 7) and the role of distributed networking technologies such as the Internet in information exchange and access (discussed in Chapter 6). In such contexts, issues of the control of information are relevant not only for understanding the political impact of information, but also its economic significance. Thus, while concerns in recent years over efforts by governments—such as China and, more recently, Iran—to impose limitations on access to information on the Internet (Jacobs, 2009a; Markoff, 2009) are related to information control (an issue discussed in more depth in Chapter 7), much of the debate over consumers' ability to make perfect copies of digital materials such as music files are equally related to information control. As such, governments can solidify their power by controlling information just as multinational corporations can stabilize their profit streams by controlling information.

As this brief discussion makes clear, simply compartmentalizing the concept of information value into two basic categories—content and control—oversimplifies the myriad ways in which information value functions, even within the specific contexts of the two categories. For instance, even the relatively simple concept of information content—what information is about—requires not only consideration of the countless different things that might make up that content but also attention to the social contexts within which information content is valued and perceived. Similarly, the concept of information control, as used in different settings, may suggest a number of different—and even contradictory—things, with a variety of implications for understanding how information is perceived and valued in those different settings.

Thus, the theory of information worlds requires a framework for understanding information value that is not limited to the dual parameters of content and control, but instead attempts to encompass the full range of factors that impact the values applied to information in different social settings. Rather than a simple duality of content and control, information value is best thought of as a spectrum—or, perhaps more accurately, a web—of interrelated issues, with the following concerns (among others) at play:

- Information Content. As noted previously, perceptions of the value of information are intrinsically tied to what that information is about and the degree to which that information meets the needs of individuals and groups.
- Perception. Information content, however pertinent in the abstract, is always mediated by normative perceptions of specific information

worlds. Within such worlds, content can be trumped by these normative pressures, while the value of information is always at least as much a matter of perception as of the intrinsic aboutness of the information itself. For instance, the same bit of information formatted and packaged in two different ways—in, for instance, a print newspaper and a blog—may be perceived differently for reasons having only to do with users' perceptions of the value of that formatting, independent of the actual content.

- Control. Because information worlds perceive the value of information differently—and often radically differently—they often can attempt to impose constraints or controls on information in order to better manage its perceived beneficial or detrimental impacts. In this sense, control can best be understood as a mechanism to maintain some kind of power over a world's ability to access information, whether the intent of imposing such control is to ensure consistent access (as in the library tradition) or to constrain access.

- Information Economies. Information, considered in terms of value, has worth, which may be measured in terms of monetary or social worth. Whether corporations profit through selling information in the marketplace or individuals profit through accessing and using the information that they need, the processes through which information is disseminated and exchanged depend upon an understanding of the worth of information. Such worth may be linked to other concepts, including perceptions of the utility of information, its profitability, or the potential of information resources to benefit some or all of the members of an information world.

This chapter attempts to explore such parameters of information value to establish a framework for many of the discussions that follow in subsequent chapters.

INFORMATION VALUE AND INFORMATION WORLDS

As information worlds occur at multiple social levels, from the purely local to the global, there exist innumerable connections across and between them. While Chatman's work focuses on small information worlds as localized communities without reference either to other small worlds or to broader social contexts, a more robust conceptualization of information worlds suggests that such small worlds do not exist in isolation, but rather are situated within—and, thus, influenced by—larger information worlds while also brushing up against other small worlds. Information worlds, thus, should be viewed not as discrete and independent worlds with impermeable boundaries between them, but rather as a continuum of social contexts, often with overlapping or fuzzy boundaries and with interactions, exchanges, and influences of various types linking them—whether tightly or loosely—to each other.

Similarly, information value is best conceived of as a continuum of attitudes and perceptions concerning the access to and the exchange and use of information across the range of social contexts, from the individual to the global, and in social, corporate, and political settings. In the most obvious sense, an individual living within the day-to-day context of a specific community will likely have a significantly different understanding of the value of information than that reflected in the policies and actions of his or her national government. Individuals and governments may not even share compatible concepts of what constitutes information in the first place, focusing on radically different things in their worlds. The individual's concerns may, for instance, be almost entirely limited to the day-to-day realities and experiences of a specific household or specific community, while the government's concerns may ignore such local concerns altogether in favor of international relations, large-scale economic patterns, and the like. Conversely, such macro-level issues may be considered to be of little interest or even utterly irrelevant by the individual, if they are even considered at all.

This apparent mutual disregard might seem to imply that different information worlds are isolated from and have limited impact on each other. Indeed, Chatman's small world theories explicitly suggest this: within a small world, according to her, "most of the information deriving from the larger outside world has little *lasting* value" (Burnett, Besant, & Chatman, 2001, p. 537). However, whether individuals residing within specific small worlds concede it, and whether they pay direct attention to the information coming into their worlds from outside or not, their local worlds do not exist in isolation, but rather are inextricably embedded within the larger social contexts of the lifeworld. The day-to-day activities of small worlds may, in some cases, unfold as if they were independent from and uninfluenced by other worlds, but pressures from outside still exist and still have an important impact on small worlds. Large-scale economic policies and laws, for instance, both constrain and enable small world activities and exert other pressures on individuals within small worlds, regardless of their day-to-day awareness or acknowledgment of such pressures. Similarly, the ubiquity of media, advertising, and other large-scale phenomena place considerable constraints on the ability of small worlds to operate as truly independent information worlds—small worlds are all subject to the influences and forces of the outside world, acknowledged or not.

Considered in relation to information value, however, the relationship between different levels of information worlds becomes a much more complex matter and is not limited simply to questions of influence or autonomy. Whether external influences and forces are openly acknowledged within individual small worlds, they exist. As a matter of information value, a world's acknowledgment—or lack thereof—of these external forces says something very important about the information value of that world.

Even though two small worlds may be similar to each other in terms of external influences and other factors—socio-economic status, proximity to a particular urban center, availability of print and broadcast media, and political jurisdiction, for example—they may differ radically in terms of the ways in which they focus on or respond to those influences. One of the two worlds may embrace the information coming in to it from external media sources and may happily accept the legal or political strictures and parameters of the jurisdictional framework within which it finds itself, while the other may largely disregard established information sources such as the media and may also choose to ignore—or even openly resist—the laws to which it is subject. Similarly, one of those worlds may perceive information primarily as a commodity which may be bought and sold and which has significant economic value, while the other may perceive it primarily as an element of free social exchange rather than as something that is linked to the marketplace and profitability.

These two worlds, then, could be said to hold significantly different information values from one another, despite whatever other similarities they may have, and each will likely hold different information values from those apparent in the larger world in which they both co-exist. Information value provides a way of talking about the differences in these worlds' perceptions not only of those things that are of importance in the world that they share, but also of the meaning and significance—whether economic, ethical, or political—of those things. The concept of information value provides a way of assessing how different worlds understand or interpret the phenomena or events with which they come into contact.

In a broad sense, information value has to do with worlds' perceptions about the significance and place of information within the boundaries of the world. Perceptions of information value determine the meaning that information has for a world, and those perceptions strongly influence what a world *does* with information—in other words, information value influences information behavior. It determines how worlds understand the worth and relevance of information content as well as how they understand the propriety of controls imposed on the information—a world's sense of information value, for instance, has a strong impact on whether they tend to emphasize open access to information or believe that certain kinds of information should be held in secrecy or be subject to legal restrictions. Further, the concept of information value provides a means of thinking about the place of information within a world in ways that are not limited to simple considerations of either control or content. A world may value a certain type of information not just because of what it is about or because it comes from an authoritative source, but because it meshes with other criteria of value—it is available in a particular medium—or because it has particular meaning in the history of the community or is freely exchanged among members of a family. Conversely, information may be rejected (as Chatman pointed out) even if it is about a critical concern or meets an important need, and even if it comes from an authoritative source, because it conflicts with a world's sense of value, propriety, and importance.

While the theory of information worlds is multi-tiered, and while information clearly can either exist simultaneously in multiple worlds or move between worlds, information value—like the other basic normative concepts derived from Chatman's work—is intrinsic to specific worlds and does not necessarily transfer easily across the boundaries between worlds. The ways in which one world values information may or may not be compatible with the values of other worlds. Thus, differences in information value between worlds, having everything to do with how they understand and ascribe worth to information, whether economically, morally, pragmatically, or otherwise, are often at the root of conflicts between worlds. Even if worlds share interests in information about specific topics, they may value that information differently and be utterly at odds with each other in terms of how they believe that information should be understood, used, or made available.

There are innumerable possible examples that could be drawn upon to illustrate such conflicts between worlds (some of which can be found elsewhere in this book, related to specific kinds of information worlds). Here, a single example must suffice. Though still using Chatman's term "worldview" rather than the term "information value," Burnett, Jaeger, and Thompson (2008) discuss a controversy surrounding the construction of the new San Francisco Public Library in 1996, focusing on issues related to the weeding of books from the collection, the amount of space in the new library devoted to computers, and the absence of traditional card catalogs in the new building. This controversy erupted when novelist Nicholson Baker criticized the library—which was an almost instant popular success in its hometown—in the *New Yorker*. Baker's vehement criticisms were met with equally vehement responses from many within the world of librarians, with the head librarian who oversaw the library's design and construction going so far as to say that Baker's claims were "bullshit, and you can quote me on that" (Kniffel, 1996, p. 13).

At its heart, this controversy involves two distinctive—and incompatible—perceptions of information value regarding what kinds of information should be held in libraries and how libraries should manage that information. The incommensurability of these two sets of information value is striking, pitting what Baker called "the old-fashioned public library of knowledge" (Baker, 1996, p. 57) against what he characterized as a market-driven, technologically mediated world with "contempt for . . . literary culture and its requirements" (p. 59). In response to Baker's outburst, a responding editorial in the pages of *Library Journal* took him to task for his "wrong" description of the library's embrace of online access to information as a dangerous for-profit "pipeline" and articulated a vision of information value that perceived the library neither as a beleaguered relic of a lost past nor as a technological wonderland (or, as Baker would have it, wasteland), but as "one stop on a *public* way" where both tradition and technology have roles to play (Berry, 1996, p. 6).

What makes this incident interesting is not simply that the two parties in the controversy disagreed about the value of a particular building, but that the two perceptions of information value, while clearly irreconcilable, also share a common goal. Both detractors and supporters of the new library would instantly agree that libraries are, as an institution, of unquestionable worth in terms of providing to their patrons access to information. Beyond that fundamental point of agreement, however, the two camps are utterly at odds in their perceptions of how libraries should go about providing such access, how they should manage their information resources, and even what kinds of information and materials should be housed within their collections.

While there are important issues at stake in an incident having to do with the truth of the two sides' perceptions (e.g., the disagreement over the San Francisco Public Library), the heart of the matter, in terms of information value, has much more to do with a somewhat different sense of truth and accuracy. The significance of the controversy depends on the degree to which each of the worlds involved perceives its own position to truly reflect the information value at stake. In the San Francisco Public Library controversy, both Nicholson Baker and the librarians with whom he does battle simply accept their standpoint as correct—and they both, even while incommensurate with each other, *are* correct, in the sense that they both accurately reflect their partisans' perceptions of information value. In the context of information worlds, this situation is a conflict between two distinct worlds and between two perceptions of information value, each of which is held to be true by the members of one world.

Thus, information value is not tied, in any simple way, to concepts such as truth or accuracy. Rather, within a given information world, the information perceived to be of value is, irreducibly, of value entirely apart from any external appraisal of its accuracy or the authority of its source. Information value is meaningful as a measure only within the context of a specific world, where it always functions in concert with the parameters, norms, and expectations of that world. Thus, conflicts between information worlds can be understood and analyzed as conflicts between groups holding often radically different understandings of information value. Two worlds can differ not only in terms of what information content they value, but also in terms of their beliefs about appropriate levels of control to be imposed on information, their assumptions about the relationship between propriety, economic value, and political goals, among many other issues. Because it touches on all of these aspects of the meaning of information within a world, the concept of information value can provide a tool for rich and robust analyses of such conflicts.

The example of the San Francisco Public Library controversy, as an event situated within a specific community, pits two small worlds against one another in a conflict that is overtly concerned with information value in all of these senses. The public arguments made by both parties not only engage questions related to what kinds of information are of value but also explicitly link that information to issues of how it should be most appropriately presented (in print or digitally), as well as to issues of the economic and political value of information within the specific social context of the library. Further, these issues are, in the controversy, fundamentally local— they reflect value as a context-specific matter, linked directly to the particulars of an information resource found in a specific world.

However, as previously noted, small worlds do not exist in isolation from the larger lifeworld within which they are situated. While the mission of the library is to serve the information needs of the San Francisco community where it finds its home, because Baker made his comments in a magazine (the *New Yorker*) named for another city a full continent distant, the conflict unfolded not only on a small local stage, but also more globally through words in national publications. The conflict simultaneously has meaning in the local context of the two small worlds engaged in it and in the broader context of the lifeworld, where it takes on significance in worlds far removed from the specific and local context of San Francisco. In information worlds (particularly in worlds permeated by mass media), what is local may also become global, and what is global may also have significance locally. When information worlds come into contact with one another, perceptions of information value are the key to how worlds interact with one another, whether cooperatively or through conflict.

As noted earlier, within a specific small world, "most of the information deriving from the larger outside world has little *lasting* value" (Burnett, Besant, & Chatman, 2001, p. 537). In the formulation of the theory of information worlds (and, in particular, of the concept of information value), the key point is somewhat different. In an information world, information deriving from outside of the world may or may not be of lasting value, but its real value—its meaning, how it is interpreted, where it is placed on a scale of importance compared to other information—is, fundamentally, local. Information may be shared across multiple worlds, but the value of that information does not pass as easily across boundaries between worlds; it is intrinsic to an individual world and is not necessarily shared—or even mutually understood—across worlds.

In this sense, when considered in relation to the concept of information value, information can be conceived of as a kind of boundary object. The same information can be available in multiple worlds, but will "have different meanings in different social worlds" (Bowker & Star, 1999, p. 297), and these different meanings are fundamentally a function of varying perceptions of information value. Like other kinds of boundary objects, information is a basis both for interaction or cross-world commerce as it is exchanged across the boundaries between worlds, and for disagreement, misunderstanding, and conflict across worlds, as the value of that information is perceived differently in each world.

TRANSFORMATIONS OF INFORMATION VALUE AND THE INTERNET

Although the Internet in terms of information worlds is discussed at length in Chapter 6, it merits some attention here as well, since its growth has important implications for the concept of information value. The remainder of this chapter will examine two examples of how information value has been transformed in the online environment: the growth of phenomena such as social tagging and folksonomies, and the challenges to the concept of rights and intellectual property in the online environment.

Bibliographic Control and Social Tagging

As noted earlier, the control of bibliographic information resources is an information value of longstanding in the library world. One of the primary tools is the use of controlled vocabularies in subject cataloging or indexing of materials. Controlled vocabularies—in which carefully constructed lists of precisely defined authorized terms comprise accepted representations of the aboutness of a set of materials being cataloged or indexed—eliminate such characteristics of human language as synonymy, homography, and other types of linguistic ambiguity in order to establish a one-to-one correspondence between concepts and vocabulary. As a result, in a controlled vocabulary, each term represents one and only one concept, and each concept is represented by one and only one term. Thus, in the service of predictable and consistent information retrieval, controlled vocabularies impose strict limitations upon the inherently ambiguous and potentially chaotic capabilities of natural human language for representing concepts.

The importance of these vocabularies in the traditional library-oriented information world can be seen in the sheer variety of examples, ranging from discipline-specific vocabularies like the ERIC Thesaurus (http://www. eric.ed.gov/ERICWebPortal/resources/html/thesaurus/about_thesaurus. html) in the field of education to general-purpose vocabularies such as the Library of Congress Subject Headings, which form part of the extensive Library of Congress Authorities (http://authorities.loc.gov/) that are used to establish authorized terms not only for subjects, but also for personal names, place names, and titles, among other entities. Historically, the significance of this kind of control can even be seen in the physicality of libraries. Prior to the large-scale advent of electronic means of housing library catalogs, the wooden card catalogs so central to library design (and so essential to the view of libraries espoused by Nicholson Baker, discussed earlier) offered an imposing embodiment of the authority and control—the *solidity*, in a very literal sense—of library values.

In recent years, the value of such control and authority has been challenged as a direct result of the Internet becoming a nearly ubiquitous presence in the industrialized world, with more than 70% of Americans online as of late 2008 and nearly 90% of respondents using search engines to track down information (Pew Internet & American Life Project, 2009). As more and more users have turned to the Internet as a primary choice among information resources, the value of library-centric structures has increasingly come to be seen by many as an anachronistic—and impractical—approach to controlling information access and representation in the wide-open and uncontrolled online world.

It is certainly not surprising, even in light of this emergence of a radically new information world, that there have been numerous attempts to apply library practices to Internet-based resources, ranging from the development of library-like classification schemes on sites like Yahoo! to the direct importation of organizational structures from the library world on sites like CyberDewey, which applies Dewey Decimal Classification coding to a relatively small set of online resources (http://www.anthus.com/CyberDewey/CyberDewey.html). However, the growth of the Internet has also brought about some interesting changes in the way vocabulary systems are conceived and developed, with important implications for understanding information value in this new environment.

Internet search engines, focusing almost exclusively on the searching of full texts rather than of surrogate records as in a library catalog, do not support the kinds of subject-oriented searches that the use of controlled vocabularies are intended to facilitate, suggesting a transformation of information value away from concepts of control and authority. However, perhaps the most interesting among such changes is the emergence of new vocabulary systems that, while still focusing on the use of discrete terms to represent aboutness, directly challenge the information value of bibliographic control that is so central to the information world of the library. Online information systems that use folksonomies and social tagging, rather than invoking the imprimatur of authority, allow users to come up with their own terms and then aggregate such user-supplied terms in a way that opens the process of vocabulary development and thoroughly does away with any formal dependence on control.

The most common implementation of social tagging has, to date, occurred in openly accessible online sites such as Flickr and YouTube (to name just two among very many), allowing users the ability both to create their own content—through, for instance, uploading photographs to Flickr—and to catalog that content by adding tags describing some aspect of the content. In addition, similar approaches have resulted in phenomena such as "social bookmarking" on sites like Delicious (http://delicious.

com/), which aggregate users' bookmarks of useful or interesting websites on a wide variety of subjects.

In each of these cases, the information value of top-down control and authority is replaced by a value that emphasizes more grassroots—or user-based—development of information, representation, and categorization. As Wikipedia puts it, information is, as opposed to control-based mechanisms favored in libraries, "generated not only by experts but also by creators and consumers of the content" (Wikipedia, 2009, n.p.). Appropriately, Wikipedia similarly treats information as something that unfolds over time through a social process involving large numbers of users rather than something subject to traditional mechanisms of control. And, indeed, Wikipedia not only treats information itself as a social process but uses that very social process as a powerful, if nontraditional, means of establishing control (Stvilia, Twidale, Smith, & Gasser, 2008).

Some critics of social tagging approaches have argued that "tagging lacks all of the benefits of controlled vocabularies. In addition, the tags assigned by some users may be so idiosyncratic or personal that they are of no real value to anyone else or may be misleading" (Taylor & Joudrey, 2009, p. 367). Others, conversely, have suggested some kind of rapprochement between the two, in which user-supplied tags augment and supplement traditional bibliographic records (e.g., Spiteri, 2006; McElfresh, 2008). While the phenomenon of social tagging clearly rests on a different perception of information value than the traditional emphasis on centralized bibliographic control, it remains to be seen to what degree the two approaches—the library world of control and the online world of social tagging—are signs of two different information worlds at war with one another or will result in some kind of hybrid approach over time.

Intellectual Property and Information Behavior

That the growth of the Internet has posed many challenges related to intellectual property is, without doubt, obvious enough to pass without extended comment here. Not only are the fundamental protocols of the World Wide Web predicated on the act of copying—in browsing, every click on a link results in material being copied from a server to the user's browser—but vast claims have been made for the autonomy of the Internet as a space apart from the strictures of existing law, where "information wants to be free." Perhaps the most notorious of such claims can be found in John Perry Barlow's (1996) "A Declaration of the Independence of Cyberspace," with its explicit claim that extant "legal concepts of property, expression, identity, movement, and context do not apply" in this new environment (see also Clarke, 2001).

Issues with intellectual property in the online environment are a clear example of large gaps between existing policies and technological realities (Travis, 2006). For example, the extensions of copyright protection to such incredible lengths—life of the author plus 80 years—create many questions of ownership, and these extensions create significant tensions with the increases in access to information brought about by the Internet and electronic files. The exceptions created to try to address these issues, such as the fair use exemption and the exemptions for use in distance education, only serve to make the issues murkier and leave many information providers and users uncertain of their legal positions (Butler, 2003; Travis, 2006). Orphan works—older works where the copyright owner is untraceable— are virtually unusable, even by the archives that own the items (Brito & Dooling, 2006; Carlson, 2005). Libraries struggle mightily with previously much clearer issues of interlibrary loan, electronic resources, and services to remote users, while universities must determine how to try to provide resources to distance education students (Allner, 2004; Carrico & Smalldon, 2004; Ferullo, 2004; Gasaway, 2000). At the same time, industries, educational institutions, and users struggle with the implications of electronic files and the ability to share files for music, movies, books, and other content formats (Strickland, 2003, 2004).

Not surprisingly, numerous legal challenges have been mounted against websites for varying forms of copyright infringement, with the battle between the music file-sharing service Napster and the Recording Industry Association of America (RIAA) being a particularly notorious example (Bowrey & Rimmer, 2005). Many such cases, like the Napster challenge, involve the use of technologies, such as peer-to-peer file-sharing networks, for unauthorized distribution of materials that are either easily digitized or are already available in digital formats (DVD and CD), such as music and film. More recently, considerable attention has shifted to print materials, which are not so easily digitized. The online availability of such materials makes a telling case study for the concept of information value. While there are numerous potential examples to choose from, in the context of information value, two challenges for intellectual property in the online environment merit specific attention: the Google Print Library Project and the more recent advent of widespread "digital piracy" of print materials.

Perhaps the most striking result of this uncertainty of intellectual property in an electronic world is the Goggle project (http://books.google.com/). The Google project began in 2004 and is conducted in collaboration with the New York Public Library, along with university libraries at Harvard, Stanford, Oxford, and Michigan. The stated goal at the outset was to digitize large numbers of out-of-print books, making them searchable and accessible online (Olsen, 2004). However, the practical result of this project is to digitize and make searchable virtually every book printed, whether or not it is still in copyright. The questionable legality of the effort has attracted some note from the mainstream press (e.g., Thompson, 2006).

In terms of perceptions of information value, such a project seems quite logical not only from the point of view of a business like Google and for many of its users who have, over time, become used to treating the search engine as a first stop for all kinds of information. However, as information value is never the exclusive domain of any given world, perceptions of information value are not uniform across multiple worlds. Unsurprisingly, the project has been consistently challenged on a number of fronts, including copyright. In September 2005, "the professional writers society The Authors Guild [sued] Google over the project, citing what it describe[d] as 'massive copyright infringement'" (Seko, 2005, n.p.). The grounds of this lawsuit explicitly raise issues of information value, in terms of both economic value (generated on Google through advertising revenue) and appropriate control over access to copyrighted materials. Most interesting, in this instance as in the case of the San Francisco Public Library presented earlier, both sides agree on a fundamental information value: for both, access to information is a positive value. Their perceptions of what constitutes an appropriate avenue for access, however, are entirely at odds: for Google, access via Google makes perfect sense; for the Authors Guild, on the other hand, "It's not up to Google or anyone other than the authors, the rightful owners of these copyrights, to decide whether and how their works will be copied" (Seko, 2005, n.p.).

The legal defenses Google has raised for digitizing and making freely available copyright-protected materials without permission of the copyright holders do not hold up well under basic legal analysis (Hanratty, 2005). Nevertheless, the lawsuit was settled late in 2008, with Google gaining the right to put out-of-print materials online and the Authors Guild gaining a $125 million fund for copyright holders (Sadun, 2008). However, Google and the Guild are not the only two information worlds with an intellectual or economic stake in digitizing print materials. A few months after the settlement, a consortium representing the information world of libraries (the American Library Association, the Association of College and Research Libraries, and the Association of Research Libraries, joined by the Electronic Frontier Foundation) asked a federal judge to provide "vigorous oversight" over the settlement, to guarantee its own vision of appropriate information value: that the subscription costs the materials housed by the Google project remain low and that the privacy of readers be maintained (Helft, 2009).

Curiously, a comment made by a reader of the *New York Times*' article on the library group's request neatly encapsulates the role of information value in this conflict. Even though Google clearly is a highly successful commercial entity and libraries have traditionally supported open and free access to information, "josh from america" (as he signs his comment) posits an information value casting Google as the champion and libraries as the enemies of access:

> Greedy ILbraries [sic]: stay away from my books.
> You had your chance, google gets books to the people freely and instantly, no more waiting for one to be returned.
> Nothing can stop the search for knowledge, don't try. (quoted in Helft, 2009)

As the comment by "josh from america" and the follow-up comments posted by others taking issue with him make obvious, information value is always a matter of perspective and perception, not fact or the perspectives or perceptions of other worlds.

Nor, in terms of information value, are information worlds necessarily monolithic entities. In the context of this discussion of the Google project, it might appear that, at the broadest level, four worlds (or perhaps five, if the comment from "josh from america" represents a definable world) have clear interests at stake: the world of Google, which is creating an information resource that will be both profitable and widely useful; the world of publishers seeking to protect profits from materials for which they own the copyright; the world represented by the Authors Guild, with an interest in defending the rights of the writers whose work might be impacted by the project; and the world of libraries, who see the project as a challenge to their historical role in providing access to out-of-print materials. The federal government may also constitute a world that has an interest in the legal settlement between Google and publishers, as the Department of Justice began in May 2009 an inquiry into possible antitrust violations raised by Google's role in and the results of the agreement (Lyons, 2009).

While many of the questions raised here will ultimately be resolved in terms of the degree to which the facts in the case mesh with the law as established by legislative bodies and interpreted by the courts, an analysis rooted in the concept of information value is less concerned with legalities than it is with the positions and points of view of the different worlds involved. None of the individual worlds involved in this situation are necessarily uniform or monolithic in terms of their perceptions of information value and how it relates to the Google project. Rather, there may well be smaller factions within each of these worlds which are, to some extent, at odds with the official positions taken by the representatives of the larger worlds providing them with their public faces. Events that unfold in the broad lifeworld may appear quite different from the perspectives of the many different small worlds that are, in one way or another, linked to the event, and even those worlds which play a visible role in the events may be less uniform than it may appear at the broad lifeworld level.

Such a variety of responses within a world that may be thought of as uniform can be seen in a more recent story in which the fundamental issues in the Google project—literary production, copyright, and online access to print materials—once again play a significant role. As reported in the *New York Times* in early May 2009, as the Google project has grown, and as digitized "e-books" have become more accessible and more widely available as authorized commodities, a "bumper crop" of illicit e-books has also appeared, freely accessible on the Web apart from the regulated and controlled environment of official publications (Rich, 2009, p. A1). It is difficult to tell from the article exactly how widespread such piracy has become or what its actual economic impact on authors and publishers may ultimately be. What is interesting about this story, from the point of view of information value, is the lack of homogeneity in the responses of writers quoted in the article.

While several—including Ursula K. Le Guin and Harlan Ellison—express grave concern, drawing a clear connection between such piracy, copyright law, and appropriate channels for providing access to their work, others offer quite different outlooks, suggesting that the small world of writers may be a composite world itself, made up of multiple small worlds, with quite different perceptions of information value. One such alternative interpretation of how book piracy impacts information value is rooted in the sense that it does not pose a real threat to the value of literary production since, as Stephen King put it, indulging in a bit of social typing to make his argument, "most of [the infringers] live in basements floored with carpeting remnants, living on Funions and discount beer" (Rich, 2009, n.p.). Another alternate perception—one that is perhaps more interesting in the context of information value—is offered by Cory Doctorow, who sees piracy not as a scourge but as an effective way of distributing his work, introducing it to a new audience: "'I really feel like my problem isn't piracy . . . It's obscurity" (Rich, 2009, n.p.).

Of course, Doctorow is not only a member of the information world of print authors but also a notable blogger (serving as co-editor of *Boing Boing*), an active advocate of changes in copyright law to reflect the realities of digital distribution and a believer in the information value of allowing work to be given away for free, in the expectation that such largesse will pay off in other ways. "The thing about an e-book is that it's a social object. It wants to be copied from friend to friend. . . . In an age of online friendship, e-books trump dead trees for word of mouth" (Doctorow, 2006, n.p.). The point here is not to argue that one or the other of these positions, as reported in the *New York Times* article, is right while the others are wrong. However, Doctorow's perception of the e-book meshes neatly with one of the fundamental arguments of this book: that the value of information is a matter of social context and is embedded within specific information worlds. Within such a framework, it does indeed seem likely that copyright law (as currently formulated and still rooted in an information world in which print rather than digital reproduction reigns) fails to acknowledge the complexity of current social context surrounding concerns of intellectual property.

Akin to copyright in terms of the challenges to established legal paradigms, another prime example of the social impacts of the Internet on information value in certain worlds can be seen in the struggle over the concept of plagiarism, particularly (though not exclusively) in the world of education, that is tied to the abundance of freely available information access online. While much attention has gone to issues like trading and distributing free electronic copies of music and books through the Internet, the capacity to find copies of papers written by others online or to buy written-to-order essays is a significant problem. This practice of simply using materials gathered online and presenting it as one's own is heavily tied to the value that many students have for online content—if it is online, it should be free and should be usable by everyone. One of the results of this, among some small worlds of students, is that "technology frames student values" (Wood, 2004, p. 237), displacing educators' quite different information values of curiosity, ethics, civic mindedness, learning, and knowledge, and contributing to a belief that anything taken from the online environment could not be considered to be plagiarized. Such is the strength of this information value that many students, when confronted with evidence of plagiarism, simply are unable to perceive it as either an infraction or a matter of educational ethics.

Ironically, this same environment serves to lessen the value accorded to certain sources of information. The Internet "makes more information available to more users, but it also makes everyone a potential publisher. A student in this environment often sees all sources as equal" (Hurlbert, Savidge, & Laudenslager, 2003, p. 41). As a result, simple ease of access can become a strong information value, according to which the information with the highest value is often the first response to a Google query. And as educators at all levels can attest, the source, quality, accuracy, or other factors by which to evaluate the information are often not applied.

The changing nature of information value in certain worlds is clearly tied to how heavily many people rely on the Internet for information and interaction and how easy it is to use, such as when one is writing a paper or looking for song. For example, people—particularly those with home access to broadband—view the Internet as a substitute for communicating with other individuals in a large number of cases, such as where an answer to a question or solution to a problem is readily sought (Wells & Rainie, 2008). As result, people often commit breaches of copyright law or educational ethics because it is easy to do and the Internet is seen as an always helpful information provider. However, the shifting values of information in some worlds may also be a manifestation of the technology giving tacit permission for a certain world to embrace an information value that the members are predisposed to, even if such values are questioned by society at large.

The 2008 presidential campaign provided a particularly dramatic example of such a situation. With a major party candidate whose name, life story, and personal appearance differed greatly from all of the major party candidates of the past, it was no surprise either that these differences were highlighted throughout the campaign or that certain information worlds looked to find negative characteristics in the candidate to justify their objection to his being different. And because he was so different, simply disagreeing with his policies was not enough for some information worlds. A number of mainstream media outlets—most notably Fox News—presented stories that served such a purpose, ranging from xenophobia (he lived outside of the country for several years as a child so he is not a U.S. citizen) to enthusiastic racism (the fist bump gesture he uses is a secret terrorist signal).

However, the majority of the stories used to justify the more inflammatory objections to this candidate gained traction on and spread across the Internet. Rumors on the Internet about Barack Obama—including that he was born in Africa, Muslim, gay, racist, and refused to say the pledge of allegiance—had a significant impact on the perception many voters had of the man, particularly in rural, predominantly white communities. In these small worlds, Internet rumors were repeated in face-to-face conversations, reinforcing the perception of their information value. These rumors were sufficiently prevalent that, at one point in the campaign, 10% of Americans believed that candidate Obama was a practicing Muslim (Saslow, 2008). Amazingly, one reporter found that the majority of the citizens of one small, predominantly white town had no problem believing the Internet-fueled and neighbor-reinforced rumors while openly acknowledging that such rumors did not match the information coming from established media sources (Saslow, 2008).

As all of these various examples demonstrate, information is always, fundamentally, embedded within particular social contexts and gains its meaning and significance—its information *value*—only within that context. The uses to which people put information—the meanings they attach to it, the degree to which they seek it out, exchange it, and derive understanding and their own kind of truth from it—are social uses and are inextricable from the information worlds, small and large, in which they live.

4 The Evolution of Information Access and Exchange

Access to information can be conceptualized in many ways; different academic disciplines have approached it in terms of knowledge, technology, communication, control, commodities, and participation, with influences on access including physical, cognitive, affective, economic, social, and political issues. LIS scholars have tended, however, to approach information access in two specific ways: as a matter of physical access, emphasizing the format and location of documents, and the degree to which users can physically acquire and use those documents; and as a matter of intellectual access, emphasizing users' cognitive abilities as well as mechanisms and structures for organizing and categorizing information. This chapter examines the roots of these conceptualizations, discusses systems such as libraries that are in place to ensure that access to information is possible, and explores the contexts that make information access so important in society.

This chapter also focuses on the concept of social access to information, a concept proposed in earlier work by the authors that serves as a third dimension for understanding access to information (Burnett, Jaeger, & Thompson, 2008). An important element of the concept of information worlds is that they are built around shared understandings of and behavioral norms regarding information; not only is information exchanged by members of a given small world as part of their ongoing social interactions, but multiple small worlds (as component parts of the larger lifeworld) intersect with each other in terms of the ways in which they understand and use information. This chapter presents information access in terms of such social aspects, arguing that information access and exchange are best understood in terms of the small- and large-scale social contexts within which they take place.

IMPORTANCE OF ACCESS

Freedom of expression is vital to democratic dialogue. However, it is not, in and of itself, sufficient to promote deliberative democracy. For democratic dialogue to have meaning, it must be based on meaningful information about the political and social issues at hand. Democracies "rest on the assumption that citizens can govern themselves because they are informed" (Hacker, 1996, p. 215). To be informed about issues, members of society must be able to have access to relevant information to inform and shape the dialogue. The availability and free flow of information "is an indispensable tool of self-governance in a democratic society" (Smolla, 1992, p. 12).

In 1969, the United States Supreme Court explicitly stated, "The Constitution protects the right to receive information and ideas" (*Stanley v. Georgia*, 1969, p. 564). Although often an unspoken element of the freedom of expression, this right to receive and to have access to information cannot be taken for granted. Without this protected right to access and to receive information, the overall right of free expression could be diminished, as could the social and political value of the speech that does occur (Jaeger & McClure, 2004; Mart, 2003; McIver, Birdsall, & Rasmussen, 2003). Reduced access to information threatens to result in a reduction in the quality and meaning of free expression and democratic dialogue.

Historian Isaiah Berlin (1996) has noted that freedom of expression does not necessarily lead to an increase in meaningful philosophical, political, or social dialogue. A reason for this seeming disconnect arises from the fact that an increase in freedom of expression is not necessarily accompanied by a similar increase in access to relevant information. "When any avenues of political expression are closed, government by consent of the governed may be foreclosed. If any information or opinion is denied expression, the formation of public policy has not been founded on a consideration of all points of view; as a result, the will of the majority cannot really be known" (Levy, 1966, p. xix). Thus, if interaction and exchange across and between multiple small worlds and across the lifeworld—all points of view—is limited or actively cut off, democratic discourse and the public sphere as a whole suffer. In other words, a deep respect for the integrity of individual small worlds—and for the information coming from them—is also necessary for the health of the lifeworld.

Consistent exposure to one viewpoint about an issue tends to limit both understanding of the issue itself and understanding of the issue within the functioning of social and political processes (Hofstetter, Barker, Smith, Zari, & Ingrassia, 1999). It is in the best interests of a healthy democracy, then, to foster the pathways through which information may be accessed and exchanged between all constituent local and national small worlds, including those reflecting diverse (and sometimes conflicting) ethnic, international, economic, religious, political, and other traditions.

Democracies, ultimately, are based on the presumption that members of society are sufficiently educated to play an intelligent role in participation and deliberation. The public sphere is devoted to these "questions of common concern" (Dahlgren, 1995, p. 7). Without access to adequate and appropriate information related to governance and other major social issues, such informed participation and deliberation are impossible. Access is a concept that can be viewed from numerous perspectives (McCreadie & Rice, 1999a, 1999b). In relation to the discussion at hand, it is useful to conceive of information access as "the presence of a robust system through which information is made available to members of society, with storage facilities ranging from physical libraries to digital databases, yoked with mechanisms for finding specific types of information stored in such facilities" (Jaeger & Burnett, 2005, p. 465). In this sense, access stands at the center of information behavior. Without access to information, there can be no exchange, use, collection, dissemination, or management of information.

The unshakable importance of access is demonstrated by the fact that it is central to the core professional statements of organizations of information professionals, such as the American Library Association (ALA), the Canadian Library Association (CLA), the Society of American Archivists (SAA), the Special Library Association (SLA), and the American Society for Information Science and Technology (ASIST), among many others. The *Code of Ethics* of the ALA, for example, proclaims "we are members of a profession explicitly committed to intellectual freedom and the freedom of access of information" (ALA, 1995, p. 1). The ALA's 1948 *Library Bill of Rights* asserts that "libraries should challenge censorship" and "cooperate with all persons and groups concerned with resisting abridgment of free expression and free access to ideas" (ALA, 1948, p. 1). To the information professions, access is central to all other information behavior.

Access can be conceptualized in many ways, and a number of different academic disciplines—LIS, communication, media studies, and economics—view access in alternate ways, approaching it in terms of knowledge, technology, communication, control, commodities, and participation, with influences on access including physical, cognitive, affective, economic, social, and political issues (Dervin, 1973; McCreadie & Rice, 1999a, 1999b). In LIS, information access has a variety of impacts on daily life. The discipline has primarily conceived of and studied information access in terms of its physical and intellectual aspects. However, information access extends far beyond these two aspects. A range of social factors can profoundly influence information access, though such an influence has not been adequately acknowledged or examined.

Within this context, information access has physical, intellectual, and social aspects, each of which can be affected by real external and internal factors, as well as by the knowledge, skills, and perceptions of individuals seeking information. An increased understanding of these different modes of information access will facilitate efforts to provide information to those who seek it. LIS researchers have long addressed information access along a number of dimensions, including, for instance, the ways in which "accessibility" is linked to personal and perceptual factors, as well as the ways in which it is a multi-faceted concept (Dervin, 1973; Fidel & Green, 2004; McCreadie & Rice, 1999a, 1999b). Studies have identified gaps between the systems-centered approach to access that focuses on information retrieval and the user-centered approach to access that focuses on information behavior (Dervin & Nilan, 1986; Julien, McKechnie, & Hart, 2005). Thus far, however, scholars have largely constrained the study of access to the physical and the intellectual, not looking to other issues that shape the process. If the study of information access is to provide realistic and inclusive perspectives, it must account for the array of social issues that significantly influence access.

In a politically and technologically complex society, the importance of information access has been institutionalized in the numerous social, political, and commercial mechanisms devoted to information access, including libraries, schools, news outlets, books, e-government, and electronic databases. Information is vital to democracy: "in the processes by which citizen preferences are formed and aggregated, in the behaviors of citizens and elites, in formal procedures of representation, in acts of governmental decision making, in the administration of laws and regulations, and in the mechanisms of accountability that freshen democracy and sustain its legitimacy" (Bimber, 2003, p. 11).

FORMS OF ACCESS

Individuals—often unconsciously, in accordance with the shared but unspoken norms and perceptions of information value embedded in their small worlds—classify information as being lesser or greater quality and value, including the "regular, repeated, familiar, quotidian, banal, and even boring" information of everyday life (Savolainen, 2008, p. 2). Such classifications are heavily influenced by the worlds to which an individual belongs and shape approaches to information access. As has been asserted previously, information access, in the context of society, is a three-fold process: physical access, intellectual access, and social access. In short, physical access is the most basic aspect—the ability to reach the information. Intellectual access is the next level of access, as it is the ability to understand the information. Finally, social access is the most advanced level of access—the ability to use the information in social contexts. All three play significant roles in information access within information worlds. While these are very abbreviated summaries of the aspects of access, these short explanations will help serve as a roadmap as the concepts are explored in greater detail later.

Physical Access

Physical access is generally viewed as access to the document or other form embodying information, be it conveyed through print, electronic, verbal, or another means of communication—literally the process of getting to the information that is being sought (Svenonius, 2000). The vast majority of discourse on information access tends to focus on physical issues, such as the physical structures that contain information, the electronic structures that contain information, and the paths that are traveled to get to information (Jaeger & Bowman, 2005).

Issues of physical access relate to the location and format of a document, and the conditions, technologies, or abilities required for reaching that document. Such issues are often readily identifiable and revolve around the questions of whether people can get into the location that houses the documents and then reach the specific documents that they seek. Physical access to information is primarily an institutional issue, depending on formalized structures that exist to ensure that the information is located somewhere and is theoretically available. This location can be physical or virtual, and the availability is not necessarily wide or egalitarian. The effectiveness of structures in facilitating the storage and retrieval of information is shaped by how well they function as intended, how easy they are to use, and how accessible they are for users with different physical abilities.

Physical access, however, also depends on knowing that the information is stored and retrievable. At the individual level, to achieve physical access the user has to know that the information exists, where it can be found, and how to navigate the institutional structures to reach it. Individual factors that can affect physical access include technology, economics, geography, and disability. Lack of necessary funds, substantial distance from or inability to use an information source, or inability to enter a location housing an information source can all create barriers to physical access. Physical access is of utmost importance; without it, no other type of access is possible (Jaeger & Bowman, 2005).

However, while it is a necessary prerequisite, mere physical access is not sufficient for full access; for instance, "it is a common, but mistaken, assumption that access to technology equals access to information" (McCreadie & Rice, 1999a, p. 51). The ability of a user to get to information and the ability of that user to employ information to accomplish particular goals are very different (Culnan, 1983, 1984, 1985). As a result, the physical aspects of information access cannot be considered without also considering the intellectual aspects.

Intellectual Access

Intellectual access can be understood as the accessing of the information itself after physical access has been obtained (Svenonius, 2000, p. 122). It revolves around the ability to understand how to get to and, in particular, how to understand the information itself once it has been physically obtained. Much less research has examined issues of intellectual access than of physical access. Issues of intellectual access involve understanding how the information is presented to people seeking information, as well as the impact of such presentation on the process of information seeking:

Intellectual access to information includes how the information is categorized, organized, displayed, and represented. Studying intellectual access can reveal the best ways to make information accessible when people act to retrieve the information and to bring the information seeker and the information together in the most efficient manner possible through representation of the available information sources. (Jaeger & Bowman, 2005, p. 67)

Intellectual access can only occur when an individual has sufficient information to engage in critical thinking and has been exposed to multiple viewpoints (Pitts & Stripling, 1990). It has been discussed in terms of a number of specific forms of information, including images, classification, catalogs and archives, government materials, periodicals, software, digital documents, and library services (Aschmann, 2002; Bednarek, 1993; Cary & Ogburn, 2000; Chen & Rasmussen, 1999; Comaromi, 1990; Deines-Jones, 1996; Dilevko & Dali, 2003; Gilliland, 1988; Intner, 1991; Mandel & Wolven, 1996; Neville & Datray, 1993; Rankin, 1992).

Intellectual access to information, at a more conceptual level, "entails equal opportunity to understand intellectual content and pathways to that content" (Jaeger & Bowman, 2005, p. 68). As a result, for an item to provide information equally to all, it must be able to produce similar outcomes or results for any user, regardless of any disadvantages that the user might have. Intellectual access to a particular item of information, however, is often not equally available.

Many personal issues unique to each user influence intellectual access, since it hinges on the user understanding the information once it has been physically accessed. Factors that can affect intellectual access can include information-seeking behaviors, language, dialect, education, literacy, technological literacy, cognitive ability, vocabulary, and social elements, like norms and values. Each of these factors has the potential to influence whether an information seeker can access the information contained in a source. Intellectual access requires the ability to understand the information in a source, which, in turn, requires the cognitive ability to understand the source, the ability to read the language and dialect in which the source is written, and the knowledge of the specific vocabulary that is used. Intellectual access also requires knowledge of the use of any necessary technology to access a source, such as telephones, computers, electronic databases, or the Internet.

Social Access

As the detailed descriptions of the concept of small worlds in earlier chapters revealed, within specific social contexts, information behavior is normative as a day-to-day activity (Burnett, Besant, & Chatman, 2001). Further, the value of information is not universal but is rooted within the norms and attitudes of a particular social world. "While one might, for instance, make use of some tidbit of information from the larger world for casual conversation with a neighbor or friend, the purpose might simply be to measure the overall soundness of the world 'out there,' to maintain a connection, or to engage in 'small-talk'" (Burnett, Besant, & Chatman, 2001, p. 537).

Although it does not address information access directly, the concept of small worlds has very important implications for information access. The social contexts of information occurring within and between small worlds interact with the physical and intellectual aspects of information access and, thus, must be taken into account in any full discussion of access. The five fundamental concepts at hand derived from small worlds (social norms, social types, information behavior, information value, and boundaries) have significant implications for information access.

Social norms allow "for standards of 'rightness' and 'wrongness' in social appearances" (Burnett, Besant, & Chatman, 2001, p. 537; Chatman, 1999) and work to establish a sense of the boundaries between a small world and the outside world around it. Information coming into a small world from beyond its boundaries that conflicts with such normative standards of propriety will seem "wrong" to the members of that world, tending to be ignored or dismissed outright as fundamentally at odds with the information value of that world. Thus, social norms may actively impact or limit information access within a small world by defining certain types of information as problematic or even dangerous.

Similarly, the information value through which communication and information exchange can—but may or may not—take place provides a constraint on what small world members are interested in or willing to pay attention to. This suggests that members of a small world will tend to consider information that does not mesh with their community's perceptions as being somehow lacking, trivial, or as something they simply do not need to know and can safely ignore. To put it in terms of social access, information value can lead a community to limit access to some information simply because it defines that information as being of little importance, regardless of its value in the world at large. However, when information has a particularly high value within a small world, especially when resources are scarce—such as among residents of a homeless shelter—members of the small world will be more reticent to exchange information with members of their own small world than with those outside the small world (Hersberger, 2002, 2003).

The concept of social types "pertains to the classification of a person or persons" within a small world. As the theory of normative behavior, upon which the theory of information worlds draws, suggests,

> if a specific individual's type is viewed as desirable within their small
> world, resources (including information) offered by that individual to

that world would be readily accepted and disseminated. However, if the individual is an undesirable type, he or she will have difficulties in overcoming this classification, and information coming from this person may not be easily accepted or believed by others. (Burnett, Besant, & Chatman, 2001, p. 537)

That is, access to important information—even information that is a matter of life and death—if it comes from somebody perceived by members of a small world as an outsider, as an unreliable source, or as somebody in conflict with the social norms or information value of their world, will tend to be limited, regardless of the content of the information and regardless of the significance of the information in the broader world outside of the community's boundaries (Chatman, 1999).

Information behavior "can be defined as a state in which one may or may not act on available or offered information" (Burnett, Besant, & Chatman, 2001, p. 537). In other words, although Chatman's theory of normative behavior applied this concept only within the constraints of specific *small* worlds, it speaks to the uses to which information is or is not put within a social world of any size, although it may have its most direct impact within specific small worlds. Information coming into one world from the outside may—if it is at odds with that world's social norms or perceptions of information value or if it comes from a source who is not trusted—be dismissed as worthless, inaccurate, or even dangerous and, thus, ignored or resisted. Conversely, information that meshes with the social norms, information value, and social types of the community will tend to be accepted by members of the world and integrated into their lives, regardless of its accuracy and value in the outside world. Still, as was suggested earlier, since every small world is embedded within larger information worlds, normative pressures from those outside worlds may still influence behaviors within small worlds, whether by placing legal or other constraints on acceptable behaviors or by impacting the kinds of information sources that are available for use.

This framework has significant implications for understanding social access to information in two primary regards. First, social norms, information value, and social types influence what information is seen to be permissible for members of a world to access and what kinds of information from the outside world will be perceived as acceptable. Second, normative information behavior defines the appropriate mechanisms and activities involved in information access, within the constraints prescribed by the various information values and social norms of the multiple levels of social information worlds. Overall, the importance of social access is often recognized, even if it is not actually articulated. For example, in 1986, an U.N. report noted, "Perhaps the most fundamental part of information is a perception" (p. 1). Thus, efforts emanating from the meso and macro levels of information worlds to impact information use and access at the micro level of small worlds must take into account not only the broad goals of providing information but also the specific norms and perceptions of information value at the local level.

Social access issues even occur in the making of information available for access. "Anyone who assigns subject categories or descriptors to a piece of recorded information is in part trying to guess the modes of access that might be found useful at some future date" (Swanson, 1979, p. 11). As such, the act of creating categories and identifiers is based on the social perceptions about access by the person making the information available and guessing how best to make it findable for those seeking access.

Studies of the acceptance of government agricultural information by rural farmers provide a telling example of physical, intellectual, and social access to information in action. Prior to the widespread usage of ICTs to distribute government information, government agencies relied on people—specially trained government employees—as the primary means of communication to specific groups. Though people are clearly constrained by limitations of knowledge, presence, physical ability, and cultural barriers, they were still the best option to get needed information to groups like rural farmers with little formal education (Woods, 1993). Studies of these programs in a number of nations demonstrated that the most effective people as communicators of government information—such as the agricultural information to help rural farmers—took into account the social considerations of the members of society they were trying to communicate with, offering practical information explained in a holistic manner that the groups could relate to based on their daily experiences (Woods, 1993).

In other words, the government agents put the new agricultural information into a form that would be understandable and acceptable to social norms, social types, information behavior, and information value of the rural farmers, leading them to accept and apply the information from someone outside of their small world and to share the new information within their small worlds and across the boundaries of their small worlds. The government agents traveling to the rural areas provided physical access, and the efforts of the government agents to provide the information within the known contexts of the farmers promoted intellectual access. Finally, by accounting for the social considerations of the rural farmers, the agents promoted social access to the information.

CONTEXTS OF ACCESS

While the examples discussed so far reveal the ways in which access affects small worlds, the combination of the impacts of access on different small worlds creates the role of access in information worlds. At the societal level, information access not only has an impact on the kinds of issues people across all kinds of information worlds focus on, but it is also a strong determinant of the type of government and the levels of freedom accorded in a society. "Since about 1776, the world has been trying to decide how much centralized-control it will allow over its affairs" (Vaidhyanathan, 2004, p. 2). When less access is granted to meaningful information for members of society, discourse in small worlds and in the lifeworld—thus across the entire spectrum of information worlds—is severely limited in its ability to consider and address significant social and political issues. In pre-Revolutionary France, for example, all printed materials had to be filtered through the King's officers and censors; political issues were not supposed to be matters for the public to consider (Darnton, 1995). In more democratic societies, however, all aspects of information access are extremely important to social and political functions. As the United States was the first modern state to formally emphasize freedom of access and expression as a right to members of society, the historical developments in the United States provide a clear story of the evolution of the contexts of information access in information worlds.

Being informed is not just an immediate concern in a democracy; it is a long-term concern. Information access serves as "a permanent record available for citizens and scholars to reflect upon indefinitely" (Quinn, 2003, p. 282). To be adequately informed about issues, members of society must be able to have access to relevant information to shape their discussions. "The state is more than an allocator of services and values; it is an apparatus for assembling and managing the political information associated with expressions of public will and with public policy" (Bimber, 2003, p. 17). By the 1810s, libraries had become official repositories for many national and state government publications, eventually leading to the creation of the Federal Depository Library Program (FDLP) to make available at specified libraries copies of all government publications (Morehead, 1999).

The United States has a long history of policies about information access; they have been key elements of the laws of the republic since its inception (Relyea, 2008). "The idea of public information was a radical concept at the time of the American Revolution" (Quinn, 2003, p. 283). The United States was based on a foundation of information policy promoting information access. The Constitution and the Bill of Rights include many key provisions related to information—the guarantees of freedom of speech, expression, and assembly, the formation of a post office, the creation of a patent office and a scheme to protect intellectual property, the preservation of religious freedom, the protection from unreasonable searches, and the freedom from self-incrimination. The creation of the national post office was particularly important to information access, allowing sources of information, such as newspapers, pamphlets, and letters—all protected by freedom of expression—to be widely disseminated.

The early political leaders clearly saw the value of widespread access to information. One of the reasons cited in the Declaration of Independence for the rebellion was the separation of meeting places of legislative bodies from the depositories of public records. The framers were careful to design the Constitution so it would function to encourage political discussion rather than foster hegemonic domination (a "tyranny of the majority") about issues (Conquest, 2000). Alexander Hamilton, James Madison, and John Jay argued in the *Federalist Papers* (1789) that the new government should be the center of information in the new nation. Information or communication appear as important concepts in 31 of the 85 essays in the *Federalist Papers*, with authors envisioning that information would both help link people to the process of governance and simultaneously prevent the formation of tyrannical majorities.

Thomas Jefferson lobbied to ensure that legislative bodies and repositories of public records be proximate to ensure ready information access for legislators (Quinn, 2003). In addition, both federal and state governments developed extensive publishing programs of government information early in the nineteenth century, with the federal government beginning to disseminate printed information to members of society in the 1810s and the Government Printing Office (GPO) opening in 1861 (Morehead, 1998). In 1895, Congress centralized all printing activities, and by 1962, all federal agencies were depositing government documents with the GPO, which now includes more than 1,300 official depositories (Barnum, 2002; Quinn, 2003). While many federal laws have limited access to very specific types of information—such as video rental records—the overall trend has been toward promoting information access (Strickland, 2005). Through the nineteenth and twentieth centuries, the progress toward making the democratic process more inclusive can be seen as paralleling increases in information access in society (Smith, 1995).

The ability to access and exchange meaningful information related to social and political issues has also been a driving force behind some of the most prominent institutions in American culture. A prime example is the public library. For many in the revolutionary generation of Americans, most notably Benjamin Franklin, diffusion of knowledge through social organizations was seen as a way to prevent the elite from having hegemony over education and learning (Gellar, 1984; Gray, 1993). Though it took many years for practice to catch up to these ideals, public libraries cemented their primary social role as the marketplace of ideas, ensuring access to materials that represented a diversity of views and interests, and opposing censorship and other social controls by the middle of the twentieth century (Gellar, 1974; Heckart, 1991; Jaeger & Burnett, 2005; Robbins, 1996; Stielow, 2001).

Ultimately, though fostering access to and exchange of diverse information sources has not always been the central social role of public libraries in the United States, it became increasingly important over the course of the twentieth century as the public library became both a place of access to meaningful social and political information and a place for members of a community to exchange viewpoints on such information (McClure & Jaeger, 2008b). As a result, libraries truly offer their societies more than information (Dowell, 2008). Perhaps the necessity of libraries to a free society is best demonstrated by the fact that while governments around the world were distracted by the Bush administration's push toward a war with Iraq, Fidel Castro ordered mass arrests of many of Cuba's librarians (Vaidhyanathan, 2004).

The vision of the framers of the U.S. Constitution for a civic republican government was "radical, egalitarian, and democratic in its implications" (Hobson, 1996, p. xi), based on a conception of government centered on "the belief that the people were the source of all authority and power" (Ferling, 2000, p. 122). This emphasis on the government working for the people, over time, has been extended to the belief that the "government does not conduct the business of the people behind closed doors" (Smolla, 1992, p. 4). Access to government information has become an essential element of democratic self-governance. In governance in the United States, it became accepted that judicial, administrative, and legislative materials and proceedings generally would be made available to the members of the public (Haiman, 1981). The unique perspective of the United States in terms of information access and exchange has led to the current situation where many nations, which otherwise have legal protections similar to those of United States, do not provide similar protections for the expression of minority or fringe viewpoints, particularly unpopular ones (Sullivan, 2001).

There have been periods of history where policy has been used to influence information behavior—particularly during the Civil War, World War I, World War II, and the Cold War (Kopel & Olson, 1996). Often, these responses include altering how information can be accessed or what information members of society can access. Responses to threats to the nation have frequently resulted in the government working to hold more information away from the public (Eberhard, 2000). The same threats have also increased uncertainty within government bodies about the best methods to restrict information access (Hogenboom, 2008). Such responses, though appealing to patriotism and a natural desire to address a threat, "create a diversion from the real issues that must be resolved" (Emerson, 1984, pp. 685–686).

There have also been sizeable changes in the ways in which ICTs have facilitated access to information. The first mass distribution of political information in the 1820s to 1830s enabled the formation of centralized political parties; the development of specialized knowledge in the 1890s to 1910s led to the creation of interest groups; and the ability to command mass national audiences through technology in the 1950s promoted the development of centralized, market-driven corporations to influence policy making in conjunction with interest groups (Bimber, 2003). In addition to making it possible for specific small worlds to establish a visible presence in the overall lifeworld, the development of the World Wide Web has also reinforced the trends that began in the 1950s with other technologies

(Bimber, 2003). These changes in the relationship between ICTs and access to political and government information have also affected government structures. The modern state relies on a centralized government authority that employs specialized information to engage in the routine functions of modern society—taxes, defense, welfare, policing, money, regulation—and these functions are greatly facilitated by ICTs (Dandeker, 1990; Giddens, 1985; Weber, 1978).

However, the general trend of U.S. history has been toward increased promotion of information access through policy (Jaeger & Burnett, 2005). Government policies that have fueled limitations on access to information have typically had negative impacts on society, from inflaming mid-twentieth-century paranoia about Communist infiltration of the U.S. government to protecting the Iran-Contra affair (Moynihan, 1998). Further, these attempts at limiting information access have not only usually become public but have frequently been halted in the face of public criticism (Foerstel, 2004; Moynihan, 1998). For example, the ALA in 1988 and 1992 published books—entitled *Less Access to Less Information by and about the U.S. Government*—to record what the ALA perceived as reductions in information access under the Reagan (1981–1989) and George H. W. Bush (1989–1993) administrations. Yet, in spite of these periods of real or perceived limitations of access to certain types of information, the general historical trajectory of information policy in the United States has been to expand access to information. The rapid expansion of the Internet as a source of information, however, raises important questions about information access in the public sphere.

THE PUBLIC SPHERE AND ACCESS

Information access stands as a vital pillar of the health of the public sphere. As the public sphere is comprised of the public spaces and forums that allow members of society the ability to critique the government and its monopoly on interpretation of political and social issues, it relies on information access to successfully serve as a channel of communication between the members of a democratic society and the political actors within the government (Habermas, 1981). The public sphere—the public press, forums, schools, libraries, and other means of free discourse about social and political information—mediate between the rights of the individual and the power of the state in democratic societies. Some entities of the public sphere, perhaps most notably public libraries, have made providing information access their defining characteristic.

In the public sphere, "there is sufficient access to information so that rational discourse and the pursuit of beneficial norms is made more likely" (Price, 1995, p. 25). Policy decisions about information—information access, freedom of expression, intellectual property, privacy, regulation of media, physical communication structures, and the support of education, research and innovation—all shape the parameters and the health of the public sphere in a society (Starr, 2004). The amount of information that is available to develop, articulate, and communicate opinions on social and political issues is key to democratic participation and to the functioning of the public sphere. The connections between information access and democratic participation can be seen as comprising three primary relationships—access to substantive information about rights and how to use them in the public sphere, access to substantive information about social and political issues for forming opinions, and channels of communication to articulate and exchange these opinions (Murdock & Golding, 1989). "Public affairs in a democracy is, among other things, a stream of collective consciousness in which certain actions are observed by politically aware citizens trying to size up events in their environment" (Mayhew, 2000, p. 5).

In the United States, an active public sphere has existed as long as the republic (Mayhew, 2000). Under policies promoting access to information and the provision of forums for discussion in the public sphere, democratic government flourished in the United States. However, that does not guarantee that such access is self-perpetuating. New ICTs can create and enhance information access in the public sphere, but they also can create new areas of contention about information access. In the mid-twentieth century, ICTs like the television and the radio were then seen as radically altering political participation in the United States. Four decades ago, it was already being observed that "the revolution in communications has indeed largely rendered obsolete . . . Madison's confidence in the dispersion of the population as an obstacle to the formation of interest groups" (Truman, 1965, p. 55). Though information-based technologies like radio and television did greatly increase avenues for information access, they did not radically alter communication between small worlds or with the government (Kakabadse, Kakebadse, & Kouzmin, 2003). In contrast, the Internet, as a communication-based technology, offers these capacities.

ICTs now allow groups to organize and communicate across an information world to argue for their agendas in the public sphere in ways that were not possible or even conceivable before the Internet. This homogenization of perspectives across an information world has the potential to increase polarization. ICTs, of course, are not alone in potentially fostering political polarization, which can also stem not only from increases in differences of opinion but also from increases of intensity of opinion, lack of historical perspective, polarization of media views, social changes, institutional changes, and even changes in social mobility (DiMaggio, Evans, & Bryson, 1997). Further, private individuals are not alone in potentially engaging in political polarization; for example, many e-government websites around the globe present information in ways that actively promote political polarization (Chadwick, 2001; Jaeger, 2005).

Polarization may, to some extent, be built into information worlds, however, since many small worlds have tendencies to isolate themselves, looking inward only toward their own localized concerns (Burnett, Besant, & Chatman, 2001). Polarization becomes a problem when external pressures encourage or even force small worlds to become more isolated in their information behavior, particularly in what they access, or when forces from large-scale and powerful information worlds such as the media or governments actively constrain the kinds of information resources available within small worlds. The ways in which information policy balances information access and regulation of ICTs, as well as social pressures like polarization, have a large impact not only on how much information is available for access but also on how democratic a society can be (Jaeger, 2007). As a result of new ICTs—especially the Internet—and regulations about those ICTs shaping information access, the nature of information access may be changing in many information worlds. In some senses, there is confusion about how governments want to handle information access and the Internet. The government of Egypt is highly oppressive, but it nevertheless is creating an archive of the entire Internet since 1986 that is publicly accessible by 200 terminals in the library of Alexandria (Vaidhyanathan, 2004).

Overall, it is believed that democracy can be enriched by the multiplication of voices so long as the mechanisms for common attention and deliberation are present (Sunstein, 2001). The Internet offers greater access to information and perspectives on meaningful issues for discussion in the public sphere. However, the same increase in the number of voices produces parallel problems about the cacophony created by all these voices, as anyone wanting to influence public discourse must compete with the millions of the other people trying to influence public discourse (Nunberg, 1996). The concerns about the creation of a forum for too many competing voices are furthered by the fact that imbalances in information serve those in power. A central dynamic of many modern states is the asymmetrical distribution of information, which is reinforced through policies that dictate access to information (Dahl, 1989).

Considering the public sphere "compels us to think about the public spaces in which new political proposals are (or should be) evaluated" (Gilman-Opalsky, 2008, p. 335). It is these public spaces that allow information to pass across boundaries and through the lifeworld while small worlds develop parallel social norms, social types, information behavior, and information value. However, depending on the ways in which they are implemented, the Internet and other new ICTs present the ability to radically redefine the public sphere and information access across information worlds. The next chapter examines a social institution that has long served as a key part of the public sphere, providing a lens through which to examine the evolution of the public sphere and its role in shaping and sustaining the information worlds of a society.

5   Public Libraries in
the Public Sphere

This chapter bridges the theoretical foundations of the book detailed in the preceding chapters and the applications of these theories in various social contexts in subsequent chapters by tracing the evolution of the public sphere as physical and virtual place through the lens of the evolution of public libraries. As long-standing pillars of the public sphere, the history of libraries through the past several centuries serves as an ideal case study of the changes in the nature of public sphere entities and their effects on information worlds, maturing into a key link between small worlds and a source of information access for the lifeworld. As many other physical public sphere entities—such as the archetypical village green—have ceased to be common the library has became the only type of physical public sphere entity commonly found in most communities. This chapter explores this evolution in relation to what the public sphere entities have viewed as their responsibilities, the social roles and expectations for these entities across information worlds, the roles of technology in the public sphere, and the migration from the public sphere as a purely physical place to a frequently virtual place.

This examination focuses on these changes not only from the perspective of members of society but also from the perspective of the individuals who work in public sphere entities. In less than a century, librarians have moved from viewing their professional mission as being moral arbiters to the dramatically different mission of being defenders of free access to all information. In the same time, the library has evolved from a physical public sphere entity where many different small worlds met and interacted to a place where patrons engage in more and more public sphere activities in virtual dispersed small worlds through the computers and Internet access provided by libraries. While libraries still serve as a hub of public sphere engagement and spreading information throughout small worlds, many librarians, curiously, resist the prominence of technology in the social roles of public libraries. All of these changes in libraries, librarianship, and technology ultimately serve to illustrate the interrelated evolutions of physical and virtual public spheres and their changing impacts on information worlds.

THE ORIGINS OF PUBLIC LIBRARIES

Libraries have existed for millennia, having gone through many permutations, functions, and levels of availability, with many early libraries having religious or scholarly functions (Jackson, 1974). As social institutions, they "have evolved in response to certain problem situations and have been shaped by countless, relatively independent individual decisions" (Swanson, 1979, p. 3). The most famous early library is the Library and Museum of Alexandria, founded circa 300 BCE by Ptolemy I, which was the center of learning in the ancient world for several hundred years (El-Abbadi, 1990).

In many subsequent centuries, libraries were confined to educational or religious settings, with the Middle Ages in Europe being the low point for libraries as scientific knowledge became equated with paganism (Manchester, 1993). After the invention of the printing press, it became possible for wealthy individuals to build personal libraries. Few of these libraries offered any opportunities for the general public to access the works, though literacy was far from common until the 1800s (Jackson, 1974). In the most dramatic sense, these libraries were intended for a very small world of users, collecting materials to fit the specific needs, norms, and tastes of the specific small group of individuals who used them.

The first popular libraries were commercial subscription libraries, formed in towns around the United Kingdom and the colonies in the early 1700s (Davies, 1974). Commercial types of libraries, such as subscription libraries and circulating libraries, tended to specialize in popular novels of the day, with the massive popularity of Samuel Richardson's 1740 novel *Pamela* driving up enrollment in many of these libraries (Jackson, 1974). At the beginning of the American Revolution, nearly a hundred libraries existed in the colonies; one hundred years later, there were more than 3,500 libraries in the United States (McMullen, 2000).

Though 1876 is considered the beginning of the modern library movement, Americans had founded thousands of libraries before then—social, circulating, subscription, academic, church, hospital, asylum, government, military, commercial, law, town, scientific, literary, philosophical society, mechanics, institute, athenaeum, and lyceum libraries, among others (Davies, 1974; Green, 2007; McMullen, 2000; Raven, 2007). Some of these were fairly well stocked with titles; for example, in 1701, the bishop of London shipped thirty-five boxes of books to Maryland for a religious library (DuMont, 1977).

American towns began passing legislation to create tax-supported school libraries in the 1830s and public use libraries in the 1840s, with states beginning to pass similar legislation a decade later (Davies, 1974; DuMont, 1977). Public libraries developed at a similar pace in the United States and the United Kingdom, with legislation for public funding of libraries becoming commonplace at nearly the same time (Conant, 1965; Davies, 1974; Gerard, 1978). Many early public libraries were established with support from philanthropists, none more prominent than Andrew Carnegie, who bestowed more than $41,000,000 to 1,420 towns to establish public libraries between 1886 and 1919 (Davies, 1974).

The intended clientele of public libraries were originally elite intelligentsia who promoted the establishment of public libraries to ensure access to the reading materials they were interested in, while secondarily serving to promote self-education of the working classes (Pungitore, 1995). The books in these libraries included dictionaries, grammars, books on political and moral issues, as well as books on practical sciences like agriculture, anatomy, astronomy, biology, chemistry, geometry, and mathematics (DuMont, 1977). As with the first libraries, early public libraries tended to focus on the information needs and values of one specific small world.

However, these narrow views of the role of libraries were not universal. From the beginning of the American republic, some leaders saw the library as a social institution that could simultaneously diffuse knowledge to members of society and prevent the wealthy and socially elite from having hegemonic domination over learning and education. Benjamin Franklin—founder of several libraries himself—was the first prominent political leader to advocate the development of libraries to provide political and educational resources to members of society (Gray, 1993; Harris, 1976). Many of the other founders of the United States, including James Madison, Thomas Jefferson, and George Mason, saw great value in official publishing of information produced by the new government and distributing these publications for easy duplication through newspapers and collection in other public institutions (Hernon, Relyea, Dugan, & Cheverie, 2002). In 1813, Congress passed the first act to ensure the dissemination of printed legislative and executive materials to selected state and university libraries and historical society libraries (Morehead, 1998).

Franklin's belief that the public library should have the primary function of promoting equality and raising the quality of national discourse was, however, somewhat unusual for several generations (Augst, 2001). Well into the twentieth century, many more civic and political leaders believed that public libraries could provide a civilizing influence on the masses and be a means to shape the populace into adhering to social norms (Augst, 2001; Harris, 1973, 1976). Libraries were "supported more or less as alternatives to taverns and streets," and librarians "viewed themselves as arbiters of morality" (Jones, 1993, p. 135). The information values and goals of public librarians were often expressed in "broadly religious terms," as if it were the library's mission to save the lost masses (Garrison, 1993, p. 37). This attitude was reflected in public library selections of materials for the public betterment and attempts to be social stewards of the general population (Augst, 2001; Harris, 1976; Heckart, 1991; Morehead, 1999; Wiegand, 1976, 1996).

The "felt cultural superiority of librarians led them to a concept of the library as a sort of benevolent school of social ethics" (Garrison, 1993,

p. 40). Andrew Carnegie's philanthropic library building activities between 1886 and 1917 further enhanced this role by making public libraries a means for improving the corporate and industrial skills of members of the public (Garrison, 1993; Van Slyck, 1995). During the period of prescriptive roles for public libraries, leaders of the library profession were greatly opposed to social change and feared the labor rights movement and other forces reshaping American society. Library leaders even generally felt that the growth of newspapers was a threat to the prescribed social order, of which public libraries considered themselves an important part (Garrison, 1993; McCrossen, 2006; Preer, 2006).

At the first ALA meeting in 1876, "most agreed that the mass reading public was generally incapable of choosing its own reading materials judiciously" (Wiegand, 1976, p. 10). Civic and political leaders believed that public libraries could provide a civilizing influence on the masses and be a means to shape the populace into adhering to hegemonic social norms (Augst, 2001; Garrison, 1993; Harris, 1973, 1976). This attitude was reflected in the elitist and paternalistic attitudes of most public libraries in selecting materials for the public betterment and in attempting to be social stewards of the general population (Augst, 2001; Heckart, 1991; Wiegand, 1996). Thus, even as libraries were becoming accessible to greater numbers of small worlds, librarians generally held to narrow understandings of information value and appropriate social norms, and attempted to enforce the information needs of a single small world across all of the information worlds in their communities.

These attitudes were already starting to change, however. Well before 1900, many city libraries had established a wide range of educational and cultural activities as a part of their regular operations, offering everything from tutoring for school children to classes teaching creative arts and practical skills for adults (Davies, 1974; DuMont, 1977). Simultaneously, libraries provided exhibits, lectures, and meeting spaces for community groups of all types and began to develop services to reach their users across multiple small worlds—such as reference services, children's services, and adult education services (DuMont, 1977). Further, libraries were also starting to model practices to improve life in communities. In cities, libraries were often among the first public institutions to adopt modern technologies and approaches for lighting, ventilation, and reducing the spread of disease (Musman, 1993).

World War I tested the burgeoning inclusiveness of public libraries, with mixed results. During World War I, public libraries actively supported the war effort by opening up facilities for use by government agencies, created war-related exhibits, promoted books about the war, served as collection agencies for bond and saving stamp drives, disseminated information provided by government agencies, promoted food conservation, and collected books to create libraries for military camps (Wiegand, 1989). They also actively engaged in censorship of their own materials—removing all kinds of German-language, pacifist, and labor-associated materials (Wiegand, 1989).

By the 1930s, public libraries more firmly began to turn away from their previous roles as agents of social control. It took the rise of fascism and a world war, but ultimately public libraries created and adopted a new primary social role as the veritable marketplace of ideas, offering materials that represented a diversity of views and interests and opposing censorship and other social controls (Gellar, 1974; Heckart, 1991). Many libraries had begun to take on a social service mission in the early 1930s, a change appearing earliest in public libraries in urban settings, where working with largely immigrant patron populations meant that active engagement with multiple small worlds became part of libraries' daily activities (Fiske, 1959). This dramatic change in social roles "emerged in an environment in which the concept of the public library's social responsibility was itself changing radically" (Gellar, 1974, p. 1367).

Numerous factors affected this reorientation, but the key change was the effect fascist governments were having on public access to information in many parts of the world in the late 1930s, specifically through lethal suppression of expression, library closings, and public book burnings (Gellar, 1984; Robbins, 1996; Stielow, 2001). In reaction to these global events, the ALA passed its *Library Bill of Rights* in 1939 and began the swing toward the modern ideal of the public library as society's marketplace of ideas (Berninghausen, 1953; Gellar, 1984; Robbins, 1996). A central component of this new stance was the unswerving assertion that voters must have access to a full range of perspectives on all significant political and social issues (Samek, 2001). Public libraries actively participated in voter registration and participation drives to increase voter turnout in the 1952 presidential election, firmly establishing the modern concept of the public library as a reliable "community source for serious, nonpartisan information on a central issue of the day" (Preer, 2008, p. 19).

After World War II, American libraries were so secure in their role in working with small worlds and promoting democracy—through supporting continuing education, serving the information needs of poor and recent immigrants, having special events for children, providing education to the working classes, opening branch libraries, and other forms of service—that a major history written at the time was proudly titled *Arsenals of a Democratic Culture* (Ditzion, 1947). However, the path to these professional commitments was not always smooth. Even after the passage of the *Library Bill of Rights* in 1939, many public libraries still banned John Steinbeck's *The Grapes of Wrath* for its political views (Samek, 2001). In the 1940s and 1950s, some public libraries were still uncomfortable with the idea of equal access to all, while others took a clear lead in the civil rights movement (Robbins, 2000, 2007).

One writer, in 1953, made clear that the need for libraries to hold to this then-new stance of providing diverse information from numerous perspectives was because "a democratic society has need for all the information it can get" (Berninghausen, 1953, p. 813). This commitment to diversity of perspectives and to battling censorship was reinforced in society when public libraries actively resisted government intrusions into library collections and patron reading habits, particularly during the McCarthy era (Jaeger & Burnett, 2005). From the perspective of the theory of information worlds, the need for diversity of information resources applies not just to the kinds of political materials typically targeted during the McCarthy era. It also covers materials more related to entertainment and other interests that may be considered by some to be of limited importance or lacking in the kinds of depth and purpose that marked the goals of many early libraries. Thus, materials including fiction and entertainment-oriented works are now important parts of library collections; the social importance of such materials can be seen in the fact that they have been challenged by those who would constrain peoples' access to information at least as often as more overtly political materials.

The first major government effort to monitor reading in libraries was an Internal Revenue Service (IRS) program encouraged by the Bureau of Alcohol, Tobacco, and Firearms (ATF) to check, without warrants, on library users who had been reading books related to explosives and guerilla warfare. The encouragement quickly spawned the FBI's infamous Library Awareness Program (LAP), a two-decade-long FBI program fishing for anything interesting in library records (Foerstel, 1991, 2004). "The Library Awareness Program was created within the FBI's bureaucracy. It was authorized by no federal law, nor did it represent any clear violation of federal law" (Foerstel, 2004, p. 35). These efforts were part of a larger FBI culture obsessed with collecting information about United States citizens, driven by the efforts of its longtime director, J. Edgar Hoover. For example, in 1921, Hoover had files on 450,000 people, an amazing feat in an era before computers; by 1974, the FBI fingerprint division had prints of 159 million people, which included most of the population of the United States (Ackerman, 2007).

Though various censorship efforts have continued to affect public libraries since the McCarthy era, such as the LAP and the USA PATRIOT Act of 2001, the public library has solidified its position as a place devoted to a diversity of ideas and small worlds, one that is open to all through its active support of widespread public access to information (Foerstel, 1991, 2004; Hartman, 2007; Jaeger & Burnett, 2005). For all these achievements, however, public libraries have difficulty expressing their actual roles in the democratic process; nevertheless, the sheer number and presence of libraries suggests that they may have greater cultural impact than librarians themselves realize (Buschman, 2007a, 2007b).

The invention of many home-use entertainment technologies led libraries to begin to include new types of media—videocassettes, CDs, DVDs, CD-ROMs—in the mission to offer users a diversity of materials with many perspectives (Pittman, 2001). As the Internet swiftly gained social prominence and significance in the 1990s, public libraries began to offer Internet access and a range of new services through numerous media that provided patrons with exposure to a wide expanse of information and ideas. Such services now function as a natural extension of the established social roles of libraries (McClure & Jaeger, 2008b). By providing new avenues by which to access information and by providing access to many materials the library could not otherwise provide for reasons of cost, space, or scarcity, the Internet can be considered a robust source of diverse, and often otherwise unavailable, information for patrons (Bennett, 2001; Kranich, 2001).

The development of the public library through the nineteenth and twentieth centuries solidified the library's social position and has created certain social roles for public libraries, maturing from a simple repository of texts to a place where a wide variety of small worlds can not only find information but also have a voice (Heckart, 1991). The idea that the public library will provide equal access to a wide range of information and views in numerous formats, often in a variety of languages representing a diverse array of perspectives on social and political issues has, thus, become a solid information value across information worlds (Jaeger & Burnett, 2005). For people with limited or no other access to published and electronic materials, the expected social function of public libraries is to ensure access to newspapers and periodicals, books of nonfiction and fiction, the Internet, music, movies, and more. The public library is now seen as a social and virtual space where all ages and walks of life can mix, exchange views, access materials, and engage in public discourse (Goulding, 2004; Jaeger & Burnett 2005). Due to their inclusive stance, public libraries are even viewed as a safe community space for members of small worlds who may otherwise feel less than accepted in general society (Rothbauer, 2007).

While it may not always work perfectly in practice, the public library remains committed to serving the needs of the full spectrum of information worlds, from the smallest of small worlds to the fullest reach of the lifeworld. Whereas many traditional public spaces in communities—the town square, the public gardens, the community market, and other places that serve to foster interaction among community members—have become less visible or ceased to exist, the public library continues to be an extremely important physical public space (Given & Leckie, 2003; Leckie & Hopkins, 2002). With so few physical manifestations of the public sphere remaining, the public library may be "the nearest thing we have . . . to an achieved public sphere" (Webster, 2002, p. 176).

LIBRARIES, TECHNOLOGY, AND THE PUBLIC SPHERE

The library is a social creation and agency that binds members of a community—with its constituent small worlds—together; thus, its roles have evolved to reflect the society it serves (Shera, 1970). The meaning of the library as a place within communities has varied across cultures, nations, and times (Buschman & Leckie, 2007). Libraries have served a range of societal needs throughout history, functioning as repository, information provider, educational institution, and social advocate (Reith, 1984). Similarly, the philosophies associated with librarianship and the principles of educating librarians have evolved over time (McChesney, 1984; Rogers, 1984). Clearly, libraries have been adaptive and changing organizations as they attempted to fill the needs of their information worlds and, in particular, the public sphere.

Technology is a significant part of the changing nature of libraries. From the time public libraries began to organize around professional associations and develop professional standards in the late 1800s, technology has been important in shaping libraries and the profession of librarianship. Melvil Dewey in particular was keenly focused on the creation and novel employment of technology to improve library operations (Garrison, 1993; Wiegand, 1996). As new means of electronic dissemination of information—such as radio, movies, and television—entered wide use, libraries reacted by eventually incorporating many of them into the services and types of materials they provided (McCrossen, 2006; Pittman, 2001; Preer, 2006). Jesse Shera (1964), in addition, foresaw that information technology could greatly reduce manual tasks performed by librarians. The progression of modern information technology has been a major influence on what libraries have tried to provide to their patrons and what patrons have in turn expected from them.

Shera (1970) asserted that technological evolution "will have tremendous importance for services which the library can offer, the ways in which it can offer these services, the advances it can make in its own technology, and in the whole underlying theory of what librarianship is" (p. 70). The specific changes eventually brought by the Internet proved, however, hard for librarians to foresee, in part because of the speed with which the Internet grew. For example, the White House Conference on Library and Information Services of 1979 anticipated the time when technology would simultaneously reduce the costs of running a library and expand the services available (Seymour, 1980). Conversely, around the same time, one library scholar asserted: "The users' perceptions of the public library, as reflected in the types of services they recognize, raise grave doubts as to the desirability of the large-scale adoption of electronic gadgetry" (DuMont, 1977, p. 128); indeed, as was noted in Chapter 3, such a suspicion of technology was still, in 1996, a central factor in the San Francisco Public Library controversy. In a collection of essays (Lancaster, 1993) written in the early 1990s by librarians envisioning the library of the twenty-first century, no mention was made of the Internet or the World Wide Web as part of the library's future, though CD-ROMs received considerable attention; while, given the stage of the Internet's evolution at that point, such an omission is perhaps not entirely surprising, it does seem, in retrospect, somewhat short-sighted. At the opposite extreme—though, perhaps, equally short-sighted—some were predicting in the 1960s that computers would wholly replace libraries before the new millennium (e.g., Kermeny, 1962; Licklider, 1965).

The rise of the Internet in public libraries has been dramatic and swift, transforming communities' expectations and perceptions of information value and behavior and radically altering what users expect from libraries, with Internet access now being an essential service for many users (McClure & Jaeger, 2008b). As Internet use by the public has increased since 1994, so too has public library Internet connectivity, jumping from 20.9% in 1994 to 99.8% in 2007, with nearly all libraries now providing public access to the Internet (Bertot, McClure, & Jaeger, 2008; McClure, Jaeger, & Bertot, 2007). In nearly three-quarters of the communities in the country, the public library is the only source of free public access to the Internet, and, thus, the only place to ensure access to such vital information and services as e-mail, online health information, online job advertisements and applications, and federal, state, and local e-government (Bertot, McClure, & Jaeger, 2008; McClure, Jaeger, & Bertot, 2007; McClure & Jaeger, 2008b).

The Internet—with its vast information resources, global interconnectivity, and means of community participation—has allowed libraries to finally achieve the vision of becoming true information centers for all needs of the community and important bridges between the local needs of small worlds and the far-flung information resources of the lifeworld. Public libraries were written about as community information centers as early as William Learned's *The American Public Library and the Diffusion of Knowledge* (1924) and *Libraries and Adult Education* (1926). The technological changes brought by the Internet have not only made the library one of the last true physical public sphere entities but also allowed libraries to serve patrons around the world through digital reference and chat services and the provision of online databases and resources. However, since the Internet has become commonplace in libraries, its true impact has been downplayed or underestimated in some quarters of the library profession, ironically paralleling how libraries initially resisted periodicals (McClure & Jaeger, 2008b).

In the early 2000s, many librarians viewed the Internet primarily as a basic reference tool that also had entertainment and communication capacities (Fourie & Dowell, 2002; Shuman, 2001). Perhaps echoing the early perceptions of the information value of libraries that saw their goal as moral arbiters and suppliers only of certain kinds of important information outlined earlier in this chapter, some have argued that the focus on the Internet displays a confusion of purpose that undercuts more traditional information provision activities (Baker, 1996, 2001; Brophy, 2007; Brown & Duguid 2002; Buschman, 2003; Tisdale, 1997). Given the mission of libraries to serve the needs—from the mundane to the essential—of multiple information worlds, however, such concerns may in fact danger-ously constrain libraries' potential.

Other fears are rooted in perception: the increasing appearance of com-puters in libraries might, some argue, make libraries appear too similar to many other social institutions. At this historical moment, the changes that libraries are undergoing make them appear to be complicit with other contemporary forces that are eroding access to history and unraveling the connections of past and future generations (Manoff, 2001, p. 374). A fur-ther factor may be that libraries have historically been considered refuges in times of social change (Rayward & Jenkins, 2007), but these social changes—in the tangible form of computers in the library buildings—reach into the essence of the library itself. "We subconsciously know that libraries are more complex than information centers" (Dowell, 2008, p. 42). For many librarians, though, the presence of computers may make libraries seem too much like information centers only. However, the theory of information worlds, particularly as it is applied in the next chapter to discuss the Internet and other information technologies, suggests that the integration of the Internet into basic library services might, in fact, serve both functions: it simultaneously situates libraries as important informa-tion centers and reflects the fact that they serve more complex needs than that would suggest.

Resistance to or fear of the Internet and its impacts are hardly unique to libraries, of course. The Internet reaches into virtually every dimension of a technologically advanced society. Computer scientist Ben Shneiderman (2008) has even suggested that the technological revolutions of the past twenty years are so all-encompassing and significant in shaping society that traditional scientific methods need to be reconceptualized. The public library is far from alone in rethinking its meaning as a social institution. Yet this issue seems to be particularly hard for certain quarters in public libraries.

Many of the problems faced by libraries in dealing with new technolo-gies and information sources through the years are similar to those faced now in relation to the Internet in that they force a reconsideration of the roles the library wants to play in society (Preer, 2006). As examples among countless others, the digital age has forced libraries to redefine the mean-ing of intellectual freedom in libraries and the meaning of the library as public forum (Dresang, 2006; Gathegi, 2005). The new technologies that have become central to librarianship have also greatly increased the ethi-cal dilemmas in providing access to information, including intellectual freedom, privacy, confidentiality, filtering, and Internet access (Alfino & Pierce, 1997; Hauptman, 2002).

Whatever the various reasons behind the resistance to Internet-enabled roles and expectations, as demonstrated previously, the public library has a long tradition of adopting new technologies to meet user information needs. In the past century, it has adopted and absorbed many different technologies to continue to expand its services, remain relevant to patrons across information worlds, and build trust in communities (Jaeger & Fleischmann, 2007; McClure & Jaeger, 2008b; McCrossen, 2006; Pittman, 2001; Preer, 2006).

LIBRARIES AND THE VIRTUAL PUBLIC SPHERE

Libraries have long served as a physical public sphere entity, with spaces for different community groups to meet and for members of different small worlds to interact and exchange information and gain exposure to new perspectives. Regardless of the technologies and types of information, libraries are still serving as a hub of public sphere engagement and spreading information throughout small worlds. Now, public libraries are "perhaps even the last true public spaces" due to their physical presence in communities and their guarantees of access to all (Leckie, 2004, p. 233). However, the Internet and related social networking technologies are also allowing the library to become a central part of the virtual public sphere, where members of society gather together in cyberspace via social networking technologies (which are discussed more fully in Chapter 6).

Instead of being a one-way information technology, the Internet features many types of multidirectional interaction and information exchange through social networking technologies, allowing for the creation of new communities online (Stephens, 2007). In June 2007, the top three Internet sites were social networking sites—YouTube, MySpace, and Facebook—attracting an astounding 350 million users that month (OCLC, 2007).

An underlying notion of these social networking applications is personal trust among participants and an appreciation of the value of receiving opinions from others (Kelton, Fleischman, & Wallace, 2008). Obtaining papers, publications, or online articles is not the same as obtaining the opinion, insights, and experiences of someone on a topic of special interest (e.g., dealing with cancer) who is trusted by the user and with whom the community of users has shared values. A major conclusion of the OCLC (2007) study *Sharing, Privacy and Trust in Our Networked World* is that users of these social sites increasingly have less concern about their privacy, confidentiality, and trustworthiness. Thus, they are increasingly likely to provide the personal information, views, and experiences that shape these sites. In doing so, they create small worlds, influence the lifeworld, and reconstruct information worlds.

These participatory technologies open up new opportunities for library services (Courtney, 2007). In the most practical sense, many libraries already offer virtual reference and chat services, online databases, and other virtual forms of traditional services. The more challenging use of social networking technologies will be to cultivate the interactions of different small worlds online and to create a meaningful virtual public sphere.

At this point, "libraries have a chance not only to improve service to their local communities, but to advance the field of participatory networks" (Lankes, Silverstein, & Nicholson, 2007, p. 32).

Libraries already play a vital role in ensuring the health of online communities—providing free Internet access to those who would not otherwise have it. Beyond providing access, the library already takes a major role in ensuring participation in some online activities. Federal, state, and local governments increasingly rely on the public library as an access point through which all members of society can reach e-government websites, with many government websites and publications even directing people to go to the public library for assistance in filing taxes, welfare requests, immigration documents, and numerous other essential government forms (Bertot, Jaeger, Langa, & McClure, 2006a, 2006b; Jaeger & Fleischmann, 2007). At just the level of local government, the most common e-government activities in public libraries include finding court proceedings, submitting local zoning board information, requesting planning permits, searching property and assessor databases, registering students in school, taking driver's education programs, applying for permits, scheduling appointments with government officials, paying fees and taxes, and completing numerous other local government functions online (Jaeger, 2009a).

A significant proportion of the U.S. population—including people who have no other means of access, people who need help using technology, and people who have lower-quality access—rely on the access and trust the assistance available in public libraries to use e-government websites (Jaeger & Fleischmann, 2007). This reliance has caused libraries to play major roles in the implementation of key policies, such as Medicare registrations and the move toward online tax payments; libraries have also played central roles in more extreme circumstances, providing access, for instance, to Federal Emergency Management Agency (FEMA) materials in the aftermath of a major disaster (Jaeger, Langa, McClure, & Bertot, 2006). In fact, the vital roles public libraries played in the aftermath of the major hurricanes of 2004 and 2005 by providing access to FEMA forms and other e-government materials essential for emergency response and recovery may have permanently cemented the public and government perception of public libraries as hubs for e-government access (Jaeger, Langa, McClure, & Bertot, 2006).

The vast majority of public libraries (78.5%) provide assistance with accessing government websites, programs, and services, while over half (55%) provide direct assistance to patrons applying for e-government services (ALA & Information Institute, 2007). Libraries, however, are not necessarily adequately preparing their own staff to meet this need. Only 27.9% of public libraries train librarians to search for and use federal government information, and 26.3% train librarians to search for and use local government information (ALA & Information Institute, 2007). Further, more advanced forms of e-government assistance, like partnering with government agencies (12.8%) and providing e-government training courses (8.4%), have not yet been widely embraced by public libraries (ALA & Information Institute, 2007). Some difficulties with new responsibilities for libraries, such as serving as the guarantor of e-government access, are not surprising, as it is yet another way libraries try to reach across information worlds to diverse users. The goals of providing service to all historically have created a great and increasing number of obligations to meet through thinly spread library resources (Conant, 1965; Gerard, 1978).

While libraries have long played the major role of connecting information worlds and creating avenues by which information can flow between and among small worlds, the biggest issue raised by ICTs in relation to libraries may be the fact that they may reduce the ability of the library to physically mix small worlds together. When members of different small worlds had to come to the physical library, they could not avoid being exposed to members of other small worlds and experiencing other aspects of the mingled information worlds around them. Now, virtual resources provided through ICTs allow users to interact without entering a physical library, eliminating the possibility of physical interactions with members of other small worlds. However, as noted previously, by virtue of the fact that in many communities they are the primary suppliers of Internet access, even the presence of Internet technology in libraries may help to mitigate this possibility, as patrons use the physical library resources to gain access to the virtual world.

Another linkage of traditional library service with new technological capacities has been intended specifically to reach otherwise underserved small worlds. Known as community-focused information services (CIS), these recent efforts focus on using new media technologies to enable users to create and share content about themselves, their small worlds, and their communities (Bishop, Bazzell, Mehra, & Smith, 2001; Durrance & Fisher, 2002; Fisher, Durrance, & Hilton, 2004). CIS projects have developed around the globe and involve not only preserving and accessing information resources but also promoting community participation and engagement (Srinivasan, 2006a, 2006b, 2007). These CIS efforts both reassert the public library as a vital community asset that can help connect patrons to locally relevant information and provide new means through which to bring library services to underserved populations within diverse communities (Becvar & Srinivasan, 2009; Boast, Bravo, & Srinivasan, 2007; Lyons, 2007; Mehra & Srinivasan, 2007; Caidi & Allard, 2005). The Chicago area provides two key examples of CIS projects. NorthStarNet (http://northstarnet.org) is designed to link people from diverse populations across the various suburbs of Chicago's urban sprawl, while SkokieNet (http://skokienet.org) not only focuses on current community members' interests—such as community revitalization, child care, jobs, and housing—but also offers to new immigrant populations that frequently move to the area resources in their native languages, including Indian, Korean, and Assyrian, among others.

Clearly, the contributions of public libraries to information worlds remain significant in the age of the Internet. Symbolically, public libraries have great social meaning as trusted social institutions that are expected to provide a range of perspectives and places for different small worlds to interact, even if they do so much less often now. A 2006 study conducted by Public Agenda found that "public libraries seem almost immune to the distrust that is associated with so many other institutions" (p. 11). Further, libraries—by ensuring information access—contribute significantly to the health of the public sphere, which contributes greatly to the overall robustness of the lifeworld. While patrons may not always rub shoulders in the physical library, they can separately use library resources to create perspectives that they can share in other forums across small worlds.

Despite these struggles with the role of technology provision and Internet access in libraries, the historical changes from library as prescriptive provider of books that reflected the norms and values of a single small world to library as the provider of a vast range of physical and virtual services attempting to reach the needs of all small worlds in a community have happened in less than a century. The evolution of the public library as perhaps the most durable agency of the public sphere encapsulates many of the rapid major social changes that have shaped the physical and virtual public spheres in recent decades, as well as the varying reactions to these changes. It also captures the rapid changes in social goals and objectives for information within information worlds.

In 1949, library researcher and educator Jesse Shera observed that "The objectives of the library are directly dependant on the objectives of society itself" (p. 248). As the social norms, social types, information value, and information behavior of small worlds are changed by technological, political, and other social influences, public libraries and other public sphere agencies can help to maintain the continuity of information worlds by ensuring that small worlds have places to connect and gain exposure to the beliefs and perceptions of different small worlds. Yet, the library also provides information to members of small worlds to help them participate in and evaluate the larger lifeworld. As the next chapter explores, these same changes in social goals and objectives are also strongly reflected in the evolution of roles of technology in information worlds.

6   Information Worlds and Technological Change

An oft-repeated claim since the earliest days of the Internet has been that it has the potential to wrest the control of information access and exchange from formal institutions such as libraries and traditional media outlets, finally truly democratizing information and putting it at the fingertips of individuals anytime, anywhere. In part, such a claim grows out of the increasingly important and increasingly public role that search engines such as Google play in the information lives of individuals using the Internet. By 2008, comments such as the following were common in the popular press, reflecting the ubiquity of the Internet in the contemporary lifeworld as well as the extent to which it has been accepted as the primary source for information in both work and day-to-day information gathering: "For me, as for others, the Net is becoming a universal medium, the conduit for most of the information that flows through my eyes and ears into my mind" (Carr, 2008, n.p.).

But the Internet has never simply been a resource for information-seeking activities. From very early on, the kinds of networking technologies that make the Internet possible have also been used to support social interaction. As early as 1973, the Community Memory project in Berkeley, California, created a proto social networking and information system with a single public terminal placed near the entrance of a local record store. "Community Memory encouraged the new. You could place your notice in the computer and wait to be instantly and precisely accessed by the person who needed it most. But it did not take Berkeley-ites long to find other uses for the terminal" (Levy, 1984, p. 148). By the mid 1980s, numerous virtual communities were in place, ranging from groups hosted by huge for-profit businesses such as Compuserve and America OnLine, through smaller pay-for-use networking systems rooted in the American counter-culture such as the WELL, to free-for-all settings such as the newsgroups available as part of Usenet. After its introduction in 1991 and opening to commercial development in 1995, the World Wide Web evolved into the *de facto* gateway to the Internet for millions of users, and untold numbers of web-based communities and discussion forums surfaced, often devoted to niche interests such as particular hobbies, television shows, or sexual practices.

More recently, considerable attention has been paid to so-called "Web 2.0" services, which actively meld information retrieval services with user-centric control and support for social interaction of various sorts. Some of these services, like MySpace, Facebook, or LinkedIn, function almost exclusively as social networking sites, defining themselves as places for friends to keep track of each other or environments for maintaining professional connections. Others, like Flickr and YouTube, operate as sites for user-supplied content such as photographs and videos. Still others, like Amazon, augment online business models with user-supplied content such as reviews and evaluations. And some take the Web's ability to support immediate updating to an extreme. Twitter, for instance, allows users to "microblog" in posts of 140 or fewer characters, letting others who follow their posts know what they are doing or thinking, moment by moment.

Increasingly, as noted in the previous chapter, libraries have integrated Web 2.0 technologies into their range of services, in an attempt to augment information access with opportunities for patron-oriented social interaction (Alexander, Carter, Chapman, Hollar, & Weatherbee, 2008; Bowman, 2008; Law, 2008; Lombardo, Mower, & McFarland, 2008). WorldCat.org, the open-access version of WorldCat, is a "global network of library-management and user-facing services built upon cooperatively-maintained databases of bibliographic and institutional metadata" hosted by the Online Computer Library Center (OCLC, 2008). It allows users not only to search its database but also to add content in the form of reviews, ratings, and notes to its bibliographic records, and even to search from their Facebook pages instead of going through the WorldCat.org website.

There has even, more recently, been talk of a Web 3.0, a hypothetical future web rooted in full integration of intelligent tools such as those that might be made possible by the development of the semantic web (Markoff, 2006). What, if anything, Web 3.0 might become remains to be seen. Fuchs (2008) projects it as the next stage of a progression that began with simple information provision in the first stage of the Web's development, followed by increased support for communication between users in Web 2.0, to a distributed setting for "networked digital technologies that support human cooperation" in Web 3.0 (p. 127).

However, some, including Tim Berners-Lee, the creator of the World Wide Web, have argued that such developments do little more than situate longstanding Internet capabilities and tendencies within a commercial context and make the connections between information exchange and social interaction explicit. As Berners-Lee has put it, even "Web 1.0 was all about connecting people. It was an interactive space . . ." (quoted in Anderson, 2006, n.p.). Still, whether such developments are truly new or simply re-position existing capabilities, one consistent thread throughout the relatively brief history of the Internet has been an important fact: current distributed networks, much more than more traditional information media, not only support access to and exchange of information but do so in a context that increasingly supports social interaction, making it at least theoretically possible for individuals and small worlds to take an active role in the production and distribution of their own information resources. It also allows small worlds the opportunity to reach across the lifeworld to connect with small worlds that would not otherwise have access to their perspectives and to create new small worlds via these virtual connections. This chapter explores some of the manifestations of those networking and information tools.

THE INTERNET AS LIFEWORLD

Historically, at least in the modern world, the availability of information across a society and to particular small worlds within the broader society has depended heavily on the presence of a robust mass media. Such media are particularly important in "large and technologically advanced countries where most of the citizenry never meet 99 percent of their fellow citizens" (Herman & McChesney, 1997, p. 2). In this scenario, a free media functions as a formalized system designed to push information out to large numbers of people at both national and local levels. While certain types of media provide mechanisms for public feedback, such as Letters to the Editor pages in newspapers and public access television outlets, they are still primarily *broadcast* media, centralized information sources rather than as open forums for social interaction.

The next chapter investigates the role of traditional media in information worlds in greater detail. However, it is important to note here that the rise of the Internet has had a significant impact on the relationship between people and the information they use, precisely because of its ability to meld information access with meaningful and extensive social interaction. While the press and other traditional media are well positioned to support the public's ability to access information—an essential prerequisite for the existence of a public sphere—the Internet adds an equally essential component: a broadly accessible virtual place where private people come together as a public to talk and collectively consider issues of public concern and other shared interests.

Such places have, of course, always been available in local family and community settings. Both the dinner table and the neighborhood pub can be venues in which social interaction, information exchange, and other public sphere activities can take place. Because it allows users to create their own virtual social spaces, the Internet has the potential to uncouple the social elements of the public sphere from the geographical constraints of the purely local, allowing groups to congregate and discuss matters of mutual interest as if they were sitting around a table together. Because it also supports the dissemination of information not only through media websites but also through blogs maintained by individuals and through even more informal channels created by groups of friends, the Internet can disengage the informational aspects of the public sphere from the broadcast oriented one-to-many model of more traditional media. Thus, the Internet has the potential to expand the availability and accessibility of information that may or may not be reported through the standard media outlets.

Perhaps because of this dual nature of the Internet as both an information source and a setting for social interaction, Habermas' notions of the lifeworld and the public sphere have frequently been used in analyses of the online world, though there has been little agreement concerning the degree to which that world truly constitutes a public sphere. Writers have, as Hess (2008) puts it, used Habermas to approach "the Internet with both hope and hesitation for its potential" (p. 37). For instance, Rheingold (1993), in his seminal and enthusiastic book on virtual communities, argues that such online settings can, even though they face the threat of commoditization, function as virtual town halls, bolstering the public sphere in the face of mass media spectacles. Similarly, Han (2008) argues that the Internet, along with other "technomedia" has some potential to support public sphere activities because it, like the print media, can foster open and rational discourse by encouraging "critical functions" (p. 29). Further, Poster (1997) uses the notion of the public sphere to suggest that a postmodern approach to politics spurred by online interaction may even transform the nature of political authority.

Conversely, others cite inequities in the patterns of links between sites, arguing that the emergence of the Internet as a public sphere domain "has not happened. Talk about 'collaborative filtering' of 'meritocracy' cannot paper over the enormous online divide whereby a few dozen educational and professional elites get more attention than the rest of the citizenry combined" (Hindman, 2008, p. 285). Wasserman (2006) argues that the Internet has been derailed by a surfeit of commercialization and triviality, undermining both its potential to provide valuable information and its potential to support meaningful interaction. "In referring to the Internet as a public sphere, one should however not lose sight of the manifold ways in which [it] provides entertainment and spectacle rather than information exchange and debate" (Wasserman, 2006, pp. 299–300).

It may be, however, that both ends of this disagreement are correct. Because it virtualizes the physical and political world in which it exists, creating an online domain that exists alongside the "meatspace" domain of the face-to-face world, the Internet tends in both directions. It is both a public sphere and a colonized space. As Fuchs (2008) puts it in a discussion of virtual communities:

> It is important to stress that [they] are not idyllic and harmonious; they are an online arena of cooperation and struggle. Characteristics of late-modern society, such as the intense colonization of the lifeworld and the whole society by economic logic, are reproduced in cyberspace; hence,

virtual communities are, besides being spaces of cooperation, also colonized by competition. Cyberspace is a contested terrain. (p. 313)

To characterize the Internet exclusively either as a public sphere domain or
as a space entirely colonized and controlled by commercial and governmental pressures is to disregard the variety of worlds and resources to be found
there. The Internet, nearly four decades into its existence (and a decade and a
half since it was opened to commercial development), cannot be simply characterized as a single, homogeneous thing. Rather, it is a complex domain
made up of smaller information worlds, a setting where "the whole ensemble
of [virtual] human relations . . . is coordinated and reproduced" through
communication practices and information exchange (Brand, 1990, p. xii). It
is a setting, in other words, for commerce as well as community, for political
propaganda as well as political activism, for idle chitchat as well as significant information exchange, for exploitation as well as cooperation, and for
education as well as the most frivolous forms of entertainment.

Within the theoretical framework of information worlds, it makes little
sense to suggest either that the Internet definitively is or is not a public
sphere space. It is important to remember that the Internet cannot be conceptualized apart from the much larger world in which it is situated and
which it mirrors. The Internet simultaneously both is and is not a public sphere—to paraphrase the American poet Walt Whitman, it is large;
it contains multitudes. And this multifarious realm can, like that larger
world, usefully be thought of as something akin to Habermas' lifeworld: a
sprawling composite of all of the information worlds and varied information resources that find their place there.

Further, like the full lifeworld of which it is part, the composite world of
the Internet contains many and varied smaller worlds, each of which, like
those worlds described by Elfreda Chatman's theories, has its own interests, concerns, social norms, and information behavior. In this sense, it
seems somewhat beside the point to dismiss the value of the Internet as an
information world simply because it "provides entertainment and spectacle
rather than information exchange and debate" (Wasserman, 2006, p. 300).
While the Internet does include "entertainment and spectacle," it is no different from the larger world in this regard; nor is the Internet, any more
than the larger world, devoid of information exchange and debate. Rather,
entertainment, spectacle, information exchange, and debate intertwine in
complex ways as components of the full spectrum of contextualized social
interaction within and across small worlds. What is considered to be trivial
or of minimal importance in one context—say, the context within which
Wasserman writes, which places particular emphasis on the kinds of rational discourse and political information also emphasized by Habermas—
can be a core value within another.

Any overview of the Internet must take this multiplicity of its worlds into
account. It could even be argued that the very makeup of the Internet as a

network of networks, with its capacity to allow the creation of dedicated spaces for individual worlds, whether with or without formal corporate and colonizing sponsorship, enforces such diversity. The longstanding cliché that the Internet "interprets censorship as damage and routes around it" (Elmer-Dewitt, Jackson, & King, 1993, n.p.) might, in this regard, be revised as "the Internet interprets colonization of the lifeworld as damage and spawns new small worlds to avoid it." These new small worlds take the form of virtual communities, blogs, wikis, other social networking sites, and a myriad of formal and informal news and information resources of varying quality. The remainder of this chapter investigates some of these small worlds that contribute to the larger Internet-enabled information worlds.

INFORMATION WORLDS AND COMMUNITY

Even before terminology like Web 2.0 was invoked to emphasize the Internet's virtual intersection between social interaction and information exchange, groups of people have gone online to find both community and information (Rheingold, 1993). While the Internet, in its earliest incarnation as the military project named ARPANet, was initially conceived of as a distributed and decentralized mechanism for sharing information, its users almost immediately commandeered its capabilities to socialize with each other, both individually via e-mail and as groups in a variety of forums including Usenet newsgroups and e-mail-based listservs and other mailing lists. As early as 1975, online communities were created for both serious purposes ("MsgGroup," the very first ARPANet mailing list, whose archives can still be found at http://web.archive.org/web/20011102110954/www.tcm.org/msggroup/) and more leisure pursuits (the Science Fiction mailing list "SF-lovers," whose archives can be found at http://www.noreascon.org/users/sflovers/u1/web/).

While both listservs and Usenet newsgroups are still common and, often, very active, many groups also use other online venues, including dedicated asynchronous Bulletin Board Systems (BBS) accessible via telephone connections, such as WELL in its early years and Echo NYC, and a wealth of both large and small real-time chat environments. With the growth of the World Wide Web, a huge assortment of text-based forums also emerged, with some older systems like the WELL also migrating in whole or in part to web-based access. Increasingly, newer community settings, such as Second Life, have appeared; they are not tied strictly to text-based interaction, but use various forms of multi-media tools to support communication and information exchange. Whatever the technology employed, these communities typically focus on a specific set of interests shared by a group of geographically dispersed participants. Such interests range from the almost absurdly trivial (such as a group of Usenet newsgroups like "alt.

swedish.chef.bork.bork.bork," devoted to mimicking the speech patterns of the Swedish chef on television's *The Muppet Show*—see their FAQ file at http://www.almac.co.uk/chef/chef/chef-faq.html) to more seriously pursued leisure activities such as soap operas or particular styles of music, to matters of life and death such as medical support groups (Baym, 1995; Burnett, 2009; Burnett & Buerkle, 2004).

A classic model for virtual communities creates some kind of discrete shared group space within which members of the community interact. Many early forms of online communities—including listservs, newsgroups, the WELL's subject-specific conferences, and web-based forums—utilize such a structure. In this model, all community interaction is shared. For instance, each subscriber to a listserv or a newsgroup has access to exactly the same set of posts made by community members; there may be closer "behind-the-scenes" connections between individual members, but a shared set of interactions remains at the core of the community per se. The focus of such communities is on the particular shared space of the community, that online place wherein open communication across the full group takes place. The social networks of such spaces unfold in public and in a way that is visible to all participants.

Some newer models of online community have transformed the ways in which these social networking links and webs are deployed across a community. While traditional communities can be said to be centered on a single commonly shared space—a particular newsgroup, for instance—networking sites such as Facebook, MySpace, and Twitter shift the focus to the individual, with each participant acting as the center of his or her own unique set of social networks. While the first model draws individual participants into shared spaces devoted to specific shared interests, a site like Facebook situates the individual as the arbiter of his or her own interests. Because a site like Facebook allows users to invite members of their own pre-existing community of friends into their networking circles, each participant's experience of the community is unique, shaped by the makeup of their own specific list of friends within the site. Further, because an individual is the center of his or her own community, and can develop and manage his or her own list of friends, there is not necessarily any definable shared subject interest across the full Facebook community. If I have a list of friends with whom I interact, that list might include family members, professional colleagues, people I know from other online community settings, high school classmates I haven't seen in thirty-five years, friends-of-friends (or even friends-of-friends-of-friends-of-friends), and so on. Other Facebook users will have their own idiosyncratic lists of friends, and each one of them provides the focal point of the Facebook community as he or she perceives it.

In a traditional online community, because of their topical focus, there is some assurance that the information shared across the community reflects a single set of norms and interests. In a setting like Facebook, on the other hand, social links cast a much wider net. A message posted to Facebook by an individual may be of interest to only some small number of his or her friends and may remain opaque to others. To put it into the terms of social network analysis, more traditional online community settings can tend to valorize community insiders through shared interactions between a somewhat persistent group of like-minded individuals over time, while sites like Facebook create more amorphous links between insiders and outsiders, often within a single group of an individual's friends.

Still, the primary defining characteristic of virtual communities is an ongoing enactment of many-to-many communication, an open process of writing, reading, and responding to texts posted in shared online spaces (Burnett, 2000). They are quintessentially *public* information worlds rooted in group-based social interaction. While individual participants certainly have their own lives and other information worlds beyond the boundaries of a specific online setting, and while individuals of a specific world may even interact with each other privately behind the scenes of the community's public space, the primary activities of the community per se occur in public through the texts posted to the community space.

Participants in virtual communities, "finding their lives touched by collectivities that have nothing to do with physical proximity" (Wilbur, 2000, p. 5), are linked through the glue of this ongoing text-based social interaction and information exchange. As component parts of the composite lifeworld of the Internet as a whole, virtual communities are almost textbook examples of small information worlds, each with its own distinct mix of socializing and information sharing, and each with its own set of social norms, social types, information value, and information behavior.

Social norms, information value, and acceptable types of information behavior are, in many communities, not only spelled out explicitly in shared documents such as FAQ (Frequently Asked Questions) guidelines or other user agreements but also emerge implicitly (though persistently) through ongoing social interaction (Burnett, Besant, & Chatman, 2001; Burnett & Bonnici, 2003; Burnett & Buerkle, 2004). Such social norms reflect standards for behaviors ranging from how participants present themselves to their virtual neighbors—whether they use "real" names or pseudonyms, for instance—to the degree to which posts are expected to be empathetic rather than antagonistic. Even across groups that might be thought of as related through similar subject interests, such social norms may vary widely, and the social norms of one may not mirror those of another. For instance, of two communities coalescing around medical issues, one may, through the preponderance of its posts, be clearly devoted to sharing accurate medical information empathetically, while the other may subordinate information exchange to the more amusing but more precarious pleasures of having participants insult each other (Burnett & Buerkle, 2004). Each of these communities melds interaction with information exchange, but each does so in its own particular way through its own information value and set of social norms.

The place of particular social types within communities is interwoven with the communities' specific emphases and normative activities and may, in fact, provide the most dramatic examples of online communities' nature as information worlds. For instance, communities that are highly tolerant of flaming, trolling, or other confrontational activities will clearly tend to be more welcoming to certain social types, preferring flamers or "trolls" to more level-headed participants. An individual thought of as a ne'er-do-well or disruptive intruder in one setting may, if particularly adept at a certain type of verbal insult or one-upmanship, turn out to be an admired figure in another (Reed, n.d.).

In virtual communities, participants interact with each other and see each other only via mediation, whether purely through the exchange of texts or through some combination of texts and graphics; the ubiquity of mediation in these settings means that the process of social typing is often visible for all to see, as an overt element of the group's interactions. In physical communities, social typing occurs primarily as a function of how community members collectively perceive an individual. Social types are most often imposed upon an individual by those around him or her, whether as a result of that individual's behavior or because of factors beyond the individual's control. In online communities, however, the mediated environment can mean that the social types of individual participants are often at least partially self-selected as part of a conscious process of self-presentation, whether through something as simple as choosing to become a member of a particular community in the first place, through the use of a pseudonym, through photographs posted to the site, through the deliberate deployment of a particular style of writing, or through choosing to discuss only certain topics at the expense of others.

Even though some aspects of social typing in online communities can be seen as willful self-presentation by individual participants, the reception of individuals by other members of a community remains a collective, shared process. Burnett and Bonnici (2003), for instance, discuss two instances in which new members entered communities and were met with very different receptions. In one, the new participant was, despite his own initial misgivings, immediately accepted as one of the group; in the other, the "newbie" clearly anticipated that she had found a welcoming group of like-minded souls but was immediately, and quite cruelly, rejected as a pretender. Why?

> First, the overall thrust of her post was, in essence, "I am lonely and I want someone to talk to," which was perceived by the group as both immature posturing and an inappropriate use of [the community's] bandwidth. Second—and, perhaps, more important—the list of interests and characteristics, even though the poster perceived them of as a

piece with the group's concerns, were perceived by the group as examples of mainstream caricature [of the group as a whole] rather than as legitimate and valid interests. In other words, through the particulars of her list, and because of certain word spellings (particularly "sooo" for "so"), the poster violated the unspoken norms of the community and, in the eyes of that community, marked herself as a pretender and outsider rather than as an authentic member of the group's culture. (Burnett & Bonnici, 2003, p. 344)

In each case, the process of social typing unfolded openly and through a combination of the self-presentation of individuals and the shared and overt interactions of the group.

Every online world, like every small world in the physical world, engages in some form of social typing, and often the markers of identity used in specific types of community can explain a great deal about how participants in the community see each other and how they perceive the parameters and value of the information exchange and other interactions within the community. Some communities place a high value on the use of creative pseudonyms as indicators of identity, while others either require or strongly recommend that users employ their actual names in interactions; some, like the WELL, combine both, allowing users tremendous leeway in the use of pseudonyms when they post messages, but also always providing links to true names. Some, like Facebook and MySpace, allow photographs of participants, other pictures, or even other graphic elements. Background images in MySpace are used as elements of users' social identities within their worlds, while others, like Twitter, use typographical conventions (such as the @ symbol) as part of their internal social conventions for users addressing each other.

Some communities even formalize the process, either valuing specific social types as important structural components of the world or using social typing explicitly as a way of defining the online identities of participants, distinct from whatever physical world identities they may have. The function of social typing as a self-conscious mechanism for presenting identity online may be most clearly seen in online role-playing and gaming communities, where social types are enforced. Participants must, in order to participate, take on the personae of fictional or imaginary characters as their "true" identities within the boundaries of the game.

Online gaming communities range from simple text-based chat environments (modeled, in many cases, on the MUDs [Multi-User Dungeon] and MOOs [MUD, Object Oriented] that became prevalent in the early 1990s) in which participants might converse with each other as if they were characters in fantasy novels like the Harry Potter series, to what are known as massively multiplayer online role-playing games (MMORPG), which, with their intensive graphics and complex social structures, can be thought of as full-fledged virtual worlds, or in the jargon of the online lifeworld, huge multi-user virtual environments.

While gaming communities are not, as information worlds, expressly devoted to the exchange of information, it does play an important role in their activities, even if it is limited to the workings and details of the game itself. In some cases, the information of value in these communities is also significant in some way outside of the community's boundaries. Participants in one of the numerous role-playing communities made up of children and young adults structured around the Harry Potter novels may, for instance, spend time discussing attempts to remove the books from library shelves, characteristics of different editions of the books, or any number of other real-world events in addition to inventing and engaging in new scenarios based on the novels' characters. Further, some games—like Sim City, a "simulation game" accessible both as an off-the-shelf standalone game and a multi-user online virtual environment—overtly attempt to mimic certain aspects of the outside world more or less realistically, although its status as a simulation is always front and center for users (Friedman, 1999).

However, in many cases, online gaming worlds maintain a strong identity purely as virtual environments, apart from the "meatspace" world in which their human participants live. The information exchanged in such worlds can be almost entirely internal to the world itself. The information exchanged by the characters within the world remains in character, focusing almost exclusively on issues within the world itself, such as laws governing character behaviors, internal economies, internal disputes, virtual world management, and the relationship between virtual worlds and the physical world in which they exist (Dibble, 1998; Malaby, 2006; Taylor, 2006).

This preference for local or internal information echoes the theory of normative behavior's suggestion that information perceived as external to a world will be thought of by participants in that world as having less lasting value than information from within the world. While gaming environments, propelled as they are by a strong element of immersive fantasy, may intensify such a tendency, it also emerges in non-gaming worlds such as Second Life. Like Sim City, Second Life mimics many aspects of the physical world, but unlike Sim City, Second Life allows users, interacting with each other via the graphic avatars that serve as their virtual embodiments, almost free rein in terms of the kinds of localized worlds they can create within its extensive virtual space. As a result, Second Life, like the Internet itself, can be seen as a virtual emulation of the lifeworld, containing many different small worlds as well as numerous larger information worlds. Many different groups, ranging from performing musicians to universities, have established a presence in Second Life for a wide range of purposes, including making or listening to music, supporting long-distance collaborations, offering virtual classroom or public lectures, mounting museum installations, making health information available, participating in sexual encounters, and even engaging in virtual marriages.

One group, the Alliance Library System based in Illinois, has led an international consortium of librarians and researchers in the creation of an "Info Island Archipelago" devoted to virtual librarianship and the provision of reference services. A year-long study of the viability of Info Island found that the vast majority of questions asked of the librarians staffing its reference desk had to do with Second Life itself, with only 186 of 6,769 questions referring to the world outside of the boundaries of Second Life (Alliance Library System, 2008). At the same time, however, a study of the Second Life "HealthInfo Island" suggests a somewhat different scenario, with an impressive amount of activity related not to virtual health but to "real-world" health issues such as HIV education and collaborations with the National Institutes of Health and the Center for Disease Control (Bell, 2008).

A general focus on matters internal to the Second Life world should not be a surprise, given the somewhat self-absorbed nature of small worlds as described by the theory of normative behavior, particularly since the long-term worth of virtual worlds in environments such as Second Life is, in the first decade of the twenty-first century, still emerging. Nor, since even in "meatspace" different worlds understand and value the same phenomena differently (Burnett, Jaeger, & Thompson, 2008), should it be a surprise that the value of Info Island or other purely virtual information services has been widely debated in library and education circles (Alliance Library System, 2008). For the discussion at hand, however, Second Life and other virtual worlds remain, at the very least, interesting extensions of other online information worlds such as virtual communities that have taken off in new directions. And it seems likely that, like earlier virtual environments, worlds like Second Life will, as they develop further, mimic the structures and patterns of more embodied information worlds. Second Life, like the "First Life" of the physical world and like the Internet itself, is an instance of a lifeworld within which a wide variety of small worlds are situated; that it is also, in turn, a small world within the larger information world suggests that it is an experiment worth continuing.

INFORMATION WORLDS AND THE PROVISION OF INFORMATION

As noted earlier in this chapter, a major thrust of online information technologies in the Internet age has been a melding of information exchange with social activities. All of the information worlds discussed so far—virtual communities, gaming environments, immersive virtual worlds, and social networking tools like Facebook and MySpace—tend to emphasize the social interaction aspects of this combination. While such worlds do, in almost all cases, place a high value on and support robust information exchange of various kinds, the role of information itself is subsumed into

the daily give-and-take of social interaction. However, rather than emphasizing the social interaction, a different set of online social information
sites place the emphasis on information itself and use various mechanisms
to incorporate interaction into their primary goal of information exchange:
blogs, wikis, and other Web 2.0 settings.

Because they place their primary emphasis on information, such sites
bear a very different relationship to the concepts of small worlds and the
lifeworld than settings that are primarily social. They are more closely akin
to the informational resources making up the lifeworld than they are to
small worlds of social environments like virtual communities.

Blogs, for instance, fall generally into three different basic categories
(Herring, Kouper, Scheidt, & Wright 2004; Herring, Scheidt, Bonus, &
Wright, 2004). The first of these takes the form of personal journals, which
function as online diaries in which an individual can offer information
about his or her own thoughts and activities to interested friends and others. Such blogs are related to more community-oriented sites such as Facebook, MySpace, and Twitter in that personal information is made available
for others in all such sites. However, diary-oriented blogs are more focused
on the individual than they are the group. Whereas a site like Facebook
displays posts by multiple people on a single page, graphically emphasizing
the social nature of the information contained in such posts, diary blogs
isolate an individual's writings into his or her own dedicated space. Such a
blog does not necessarily rely on the web of social connections, but it can
be thought of as a standalone record of personal information, even if it is
meant to be read, like a diary, only by its creator (Crumlish, 2004).

The second type of blogs, identified by Herring, Scheidt, Bonus, & Wright
(2004) as knowledge logs or k logs, are less personal in nature, functioning as repositories of information about a particular topic, often related to
information technology. As repositories, such blogs can be thought of as
information resources offered up by experts, actual or sometimes self-pro-
claimed, for the use of interested parties or other experts, which, because
they are created by individuals, play an important role in the social worlds
of their creators and readers.

Blogs in the third general type, filter blogs, spotlight some aspect of the
world at large of interest to the individual (or small group of people) who
maintain them, often focusing on politics or other sorts of current events.
This category of blog includes such high-profile sites as *Boing Boing* and
Daily Kos: State of the Nation. Some of these blogs are maintained in conjunction with major traditional media sources, while others, like *Boing
Boing* and *Daily Kos*, are independent.

Most blogs—and even many current newspapers, whether or not they
are actually structured as blogs in their online editions—support readers'
ability to add their own comments, and some have extremely active commenting areas as well as large groups of enthusiastic or critical followers
(as can be seen in the example of josh from america's comments to the *New*

York Times cited in Chapter 3). They generally function as gatherings of information of one kind or another, chosen by the blogger as a reflection of personal interest. Blogs have been widely studied as a particular type of virtual community, complete with its own set of small world social norms and information value and yoked together through blog-to-blog links, reciprocal citation, and the practice of give-and-take commenting across linked blogs (Blood, 2000; Crumlish, 2004; Miller & Shepherd, 2004; Nardi, Schiano, & Gumbrecht, 2004; Viegas, 2005; Wei, 2004). Although the "blogosphere" can be seen broadly as a setting for community in a couple of different ways—through interconnected, mutually linked blog sites and through the ongoing feedback of reader comments—they more importantly function as mechanisms for the exchange of information outside of more traditional media channels.

While they may often lack the formal imprimatur of recognized media outlets, and while there may be concerns in some cases about the authority or accuracy of the reporting of bloggers, the very existence of blogs opens up the potential for public sphere activities. The presence of multiple bloggers working independently of media- and government-dominated information channels can help to ensure that important information that might otherwise be suppressed or downplayed remains in circulation and accessible (Braman, 2006). Similarly, bloggers, independently of the accuracy or authority of the information they include in their blogs, may serve as gatekeepers between their own particular small worlds and the larger lifeworld. Members of the small worlds of regular blog readers largely view them as credible sources of information (Johnson & Kaye, 2004). Conversely, they may also play important roles in making information of value within particular small worlds available beyond the boundaries of the small world, thus further enriching the lifeworld.

Blogs, largely the productions of interested individuals, can be thought of as examples of ongoing independent reportage, of individual and often idiosyncratic offerings of information from a particular vantage point (Crumlish, 2004). The social facet of blogging surfaces not so much in the production or presentation of the information they contain, but rather as a function of the larger setting within which individual posts appear. Posts offer up information to individuals other than the blogger, who subsequently have the option of responding through comments or through adding links from their own blog to the post.

By comparison, wikis can be thought of as small- or large-scale information resources expressly created through explicit open social processes. Wikis are collaboratively authored websites that allow readers to add new content or to edit existing content in an ongoing process of creation and collaboration, constantly shifting the website landscape (Leuf & Cunningham, 2001). For the most part, wikis maintain an ongoing and openly accessible record of edits and changes, along with a forum within which participants can discuss such changes. Thus, while blogs are social information sites by virtue of their larger context and one part of their structure (i.e., because of the interlinkings across the blogosphere and because of the presence of reader comments), wikis are social information sites by their very nature. Wikis expressly situate the production of information resources within an overt and intentional set of social processes. As instances of information worlds, wikis are particularly interesting insofar as they are simultaneously clear examples of both the promise and the problems inherent in online social information exchange and the interactions between small worlds and the lifeworld in online settings.

In recent years, much attention has been focused on Wikipedia, the online "free encyclopedia" which is the most notorious and almost certainly the largest wiki, with versions in multiple languages, nearly three million articles in English (as of early 2009), and hundreds of millions of visitors each year. Much of the attention paid to Wikipedia has emphasized the quality of its information and its relative accuracy when compared to more traditional encyclopedic resources such as the *Encyclopedia Britannica* (Giles, 2005; Stvilia, Twidale, Smith, & Gasser, 2008). Since its creation in 2001, Wikipedia has clearly become a default first-stop information resource for many people, both by virtue of the extensive body of information it makes available on its own and because Google often places Wikipedia resources at or near the top of its display of search results.

While some have argued that both it and Google play significant roles in perpetuating the colonizing tendencies of the "'same old' media biases of mainstreaming, hypercommercialism, and industry consolidation" (Diaz, 2008, p. 11), Wikipedia has clearly become a de facto major resource for the lifeworld of the Internet. Further, because of the way Wikipedia articles are open to ongoing edits by readers, it can also reflect the interests and perspectives of particular small worlds. Indeed, Wikipedia, which has been called "an 'impossible' public good" (Ciffolilli, 2003, n.p.), makes such a link to the information worlds both small and large explicit, presenting itself not just as an information resource but also as a "social community" of writers and editors (Wikipedia, 2006a). Perhaps the most direct relevance of Wikipedia to disparate sets of information worlds can be seen in its ongoing battles over content. In numerous instances, individuals (including congressional staffers) have edited (or, from another point of view, *vandalized*) pages to reflect either their own individual interests or, more commonly, the perceptions and values of the small worlds of which they are members (Wikipedia, 2006b). Entries, both significant and trivial, have been subjected to *edit wars* over matters ranging from factual substance to spelling and other editorial standards; in such wars, two (or more) different groups engage in repeated back-and-forth edits to make a particular entry reflect their own small worlds' perspectives (Wikipedia, 2006c). Wikipedia's own sense of information value can be seen in the openness with which it allows such controversies to unfold, maintaining a record not only of every edit, but also discussion forums in which they can be discussed by those interested.

While such occurrences have important implications for the long-term potential of Wikipedia and similar sites as pure information resources, from the perspective of an analysis rooted in the theory of information worlds, they are important because of the clear ways in which they exemplify the intersections between social activities and information provision, as well as the interactions between small worlds and the larger lifeworld and across information worlds. Projects like Wikipedia, thus, can be seen as important attempts to reconcile the different interests and agendas of these multiple and intersecting worlds. Given the persistence and vehemence of edit wars, the degree to which Wikipedia continues to be, in fact, a quite reliable—not to mention popular—information source suggests the potential of such attempts to meld social interaction and information provision to meet the needs of both individual small worlds and the lifeworld as a whole.

While Wikipedia is a non-profit project, the approach of allowing users to manipulate and/or create information has also made an impact on the online marketplace, particularly under the aegis of Web 2.0 applications. While the availability of not only customer reviews but also extensive discussion forums does not make Amazon either a virtual community or a small-world-based information resource, it does at least have the potential of adding the voices of consumers to the profit motive of the marketplace. It would be a mistake to argue either that such a mix of marketplace economies with user interaction is solely an instance of the public sphere or that it is solely an example of the colonizing of the lifeworld by corporate interests. Rather, it seems likely that it, like the Internet itself, includes elements of both. Neither entirely fish nor fowl, the types of Internet phenomena explored in this chapter remain hybrids of various sorts, melding small worlds with the lifeworld, information provision with social interaction, and the public sphere with its own antithesis.

With the online universe still in relative infancy (Markoff, 2009), whether it will ultimately tend to enable or limit the voices of different information worlds remains to be seen. But, at this stage of its evolution, it is at the very least a significant and robust setting for a rich variety of interwoven information worlds. Individuals can use online social networks to become exposed to the social norms, social types, information behavior, and information value of small worlds that they would not otherwise encounter. Simultaneously, small worlds can use these avenues to try to shape perceptions in the lifeworld by broadcasting their social norms, social types, information behavior, and information value. Ultimately, the Internet and its capacities radically shift the function of boundaries between small worlds, as small worlds are no longer limited to bounding other small worlds in physical sense, boundaries between worlds can also be virtual. As the next chapter will explore, the influences of small worlds on information worlds can occur not only through the channels of the Internet and its social networks but also through what is regarded as mainstream or traditional media.

7 News, Media, and Information Worlds

Early in the history of the American Republic, James Madison, the most significant contributor to the development of the constitutional government wrote, "A popular Government without popular information or the means of acquiring it, is but a Prologue to a Farce or a Tragedy or perhaps both. Knowledge will forever govern ignorance, and a people who mean to be their own Governors, must arm themselves with the power knowledge gives." While this insight has been true throughout the history of all modern democracies, the shifting goals and roles of media in the United States and other industrialized nations demonstrate how the types and amounts of information provided throughout the media can shape discourse in information worlds at all levels. Generally, "the more democratic a given society, the more publicly accountable would be its broadcasting system" (Herman & McChesney, 1997, p. 14).

As independent media presenting varying perspectives on important social and political issues has long been held to be necessary for a functional democracy, the historical development and current state of media is core to understanding information behavior in information worlds of all sizes. "[A]ll new media emerge into and help reconstruct publics and public life" with "broad implications for the operation of public memory, its mode and substance" (Gitelman, 2006, p. 26). The amount and quality of information available across information worlds is heavily dependant on the mass media in a society. As the term "mass" denotes, it is the role of the media to broadcast information as widely as possible, touching on as many worlds as it can reach. Historically, the formats and methods of information dissemination have evolved, but, as has been previously detailed, the importance led Habermas to suggest that the public sphere was initially impossible without an active and vigorous media. The public sphere, in his view, was to provide "critical regulation" of both the government and society (Warner, 1993, p. 8).

In democratic societies, the media has played many different roles, but regardless of the nature of such roles, media has been long considered a vital part of an informed democratic population, particularly one in which the population is large, diverse, and/or geographically dispersed. The content of media—both mainstream and non-traditional—heavily influences discourse both across the lifeworld and within individual small worlds by making information available, by presenting it in a certain manner, and by choosing not to present other information at all. Modern corporate media, a clear information world with its own political and commercial interests, has very specific agendas in the presentation and packaging of information.

Within a society, different small worlds will have varying relationships with the media—some reject it entirely in favor of the opinions of people they know, some consume it without thought, and most only pay attention to media outlets that suit their personal views. However, with growing corporate and government influence over media content over the past several decades, the views available in mass print, broadcast, and electronic media have actually constricted and become homogenized as sources of information, although increases in the number and range of non-mass-media sources such as blogs may work to mitigate this constriction somewhat. Still, as has been argued earlier, even the proliferation of small world information sources through the blogosphere and other online settings might be seen as a kind of double-edged sword. While the online environment, as a type of media, offers many new sources of information and the means of creating new small worlds through which individuals can exchange information, it might also, because of the way it can present information tailor-made to specific interests and small world norms and values, provide inhabitants of those worlds only the information that they want to see, presenting avenues by which they can become more insulated from the larger information world and can gravitate toward more extreme positions. Such polarization can also be encouraged by the members of the powerful information worlds who control the media, when they choose to use their power intentionally to shape perceptions both within small worlds and across the lifeworld.

MEDIA AND DEMOCRATIC GOVERNANCE

In the United States, the framers of the Constitution envisioned a tightly interwoven fabric of information worlds, working together to ensure the health of the nation. As an important element of this social fabric, the press would provide the most important check and balance to executive power, perhaps most importantly when executive power swelled in times of crisis, such as war (Nichols & McChesney, 2005). Other democracies have since followed this general notion, while despotic governments have applied the inverse of the same lesson by severely limiting or completely controlling the media. It has been aptly noted that power and knowledge are directly interrelated, with power able to produce or limit knowledge (Foucault, 1979). "Access to information has always been a condition of social power, and the economy of even the most primitive societies must have been to a significant extent dependent on information flows" (Duff, 2000, p. 172). The modern conception of the information world of the media—corporate-owned, for-profit news divisions staffed by professional reporters—has not always defined the role of the media in society. In fact, the framers would not have even understood the term "the media."

At the time of the adoption of the Constitution, the press in both the United States and the United Kingdom were primarily comprised of local-ized small worlds, printing and distributing local papers and broadsides. The value accorded to these papers can be seen in a 1791 article by James Madison, who wrote "Whatever facilitates a general intercourse of senti-ments, as good roads, domestic commerce, a free press, and particularly *a circulation of newspapers through the entire body of the people*, is equiva-lent to a contraction of territorial limits and is favorable to liberty" (empha-sis in original). While each of these small world outlets usually had a very distinct point of view on the issues of the day, the contrasting voices were generally completely independent and driven by the perspectives of the edi-tor of the paper. These early newspaper writers were "fierce partisans and some were talented party organizers, but none of these journalists were simply party men" (Daniel, 2009, p. 6). These small worlds, with their contrasting views were, to Habermas, essential to the functioning of the public sphere and the wider lifeworld—many independent voices vigor-ously debating the most important social and political issues of the day. In the United States, the guarantees of freedom of the press and freedom of speech, as well as the development of a national postal system, ensured an environment in which papers could print their opinions and have them distributed around the country, thus benefiting small worlds and the life-world alike.

The press of the early 1800s could be highly political, with the presi-dential contest between John Adams and Thomas Jefferson in 1804 being a strong example of the freedom of the press being used by independent newspaper editors to heatedly argue for highly political goals (Ellis, 1996; McCullough, 2001). However, the political debates carried out partly through the newspapers typically led to compromises on the issues at hand, as creating a livable consensus was a standard goal of political debate at the time (West, 1997). These kinds of partisan journalism helped to con-textualize issues so that citizens in particular small worlds could recognize patterns in events and understand broader social trends in the develop-ing society (Nichols & McChesney, 2005). This approach to journalism was heavily influenced by the political philosophies of the early American republic. In debate and discourse, the members of the founding genera-tion used "party" and "faction" interchangeably because political parties were seen initially as a means for people with different perspectives on an issue to organize their voice on a specific issue (Truman, 1971). However, attempts by modern media companies to influence events in order to sell their products are far from new (Baldasty, 1993). The "entrepreneurial view of journalism had given rise to the penny press of the 1830s," though this style was far from dominant until the late 1800s (Baldasty, 1993, p. 99).

Technological change was a prominent driver of the transition from the press as a myriad of small worlds benefiting the health of the public sphere to the corporate media, as it allowed the newspapers to play a larger role in how government responded to events. Mechanization of printing, cutting, and folding of papers in the mid-1800s led to enormous increases in distribution and readership, as did increasing levels of literacy (Hanson, 2008). Soon after, the telegraph increased the power of the press by allowing them to report on events soon after they occurred and spread these reports around the world (Hanson, 2008). The increased speed of the spread of news forced governments to react to events at a much faster pace. "The telegraph was the first technology to inform the public as well as leaders about events as they were still occurring. There was less time to make decisions. The press gained influence. Public opinion could be more easily aroused" (Hanson, 2008, p. 20).

As the telegraph was giving the press more power, it was able to spread stories at a much faster rate nationally and internationally, simultaneously affecting the role of the press in local small worlds and extending the reach of the lifeworld. The volcanic explosion and subsequent tsunami from the island of Krakatoa in 1883 was the first major news event spread worldwide via the telegraph (Winchester, 2003). This power to spread information around the globe also led many news organizations to try to influence news directly to sell papers, going so far as to try to manufacture conflicts to build readership. Some newspapers would take these efforts too far, however, going to the extreme of fostering the Spanish-American War.

The concept of professional journalism was launched in the early 1900s, as the sensationalism and right-wing propaganda that dominated news had begun to negatively impact profits and three of the four major presidential candidates in 1912 attacked the corruption of the press (Nichols & McChesney, 2005). Professional journalism was sold to the government and to the public as a break away from using the news to sell partisan interests of press owners (Nichols & McChesney, 2005). Professional journalism, however, in part because it allowed the "news" to be perceived as the same thing across many small worlds, also helped speed the process of macro-level corporate consolidation, profit-driven news organizations, and limitations of the approaches to news gatherings.

In the United States, the development of corporately controlled commercial media—network-centered and advertising-supported—has been a long process, but the dominance of commercial media was established by the late 1930s (McChesney, 1993). With consolidations of first newspapers and radio, then television and eventually Internet-based channels, commercial media became cemented as a major social force. The main period for attempts to reform and democratize broadcasting in the United States was 1930–1933 (McChesney, 1993).

World War II brought broadcast media into the realm of armed conflict. While broadcasting had been used in the 1920s and 1930s as a means of propaganda and control of empire by a number of governments, it quickly became a weapon of conquest and war during World War II, with many nations using radio to bolster their own populace and to try to intimidate their adversaries, either emphasizing one nation's social norms and particular set of political beliefs and information values or assailing another's by broadcasting both at home and abroad (Wood, 1992). For example, prior to America's entry into World War II, German propaganda broadcasts to the United States emphasized creating anti-British sentiment rather than promoting pro-German sentiment, but this approach was changed as soon as America entered the war (Chester, 1969). These uses—with their very particular vision of information value—have continued to be common since, with "information, misinformation, mirror broadcasting, psychological warfare and covert broadcasting" being used to "inform and deceive, justify and confuse, disorient and bring about internal uprising, sabotage, and terror in the enemy country" (Wood, 1992, p. 243).

The Voice of America (VOA) was established to disseminate a U.S. perspective on world events and familiarize people in other nations with U.S. cultural and political life (Puddington, 2000). During the Cold War, Radio Free Europe and Radio Liberty encompassed more than twenty different stations providing content directed at each member country of the Soviet Union, acting as an alternative to state-controlled media by offering news, features, and other content of specific interest to residents of each of the Soviet satellites (Puddington, 2000). Many other nations developed broadcast systems similar to VOA—including Australia, Canada, Cambodia, China, Croatia, Egypt, France, Germany, Iran, Korea, Kuwait, Laos, Libya, Myanmar, the Netherlands, Russia, Slovenia, Sweden, Taiwan, and Vietnam, among others—intended to spread information, disseminate propaganda, and influence regional and world opinion (Wood, 2000).

Whether they traffic in the dissemination of accurate information or propaganda, outlets like the VOA and similar efforts can be seen as attempts to inject the norms, beliefs, and information value of one information world into others. They are cases of the government using government-controlled media to convey its messages and accomplish its objectives. However, in the 1940s, not only had commercial media come to dominate the news channels available to support the public sphere, the government discovered that it could directly influence what was presented to small worlds within and throughout the country by creating its own programming. During the 1940s and 50s, more than twenty-five news and public affairs series that consisted entirely of programming provided by the federal government aired on the networks, with the military providing hundreds of additional films for stations to broadcast (Bernhard, 1993).

This relationship between commercial media and government grew increasingly close, often resulting in journalists moving far away from the goal of providing information either for specific small worlds or for discussion in the public sphere. During the Korean War, to avoid displeasing sponsors or government officials, the networks requested that the government formally impose censorship on reporters, resulting in guidelines that prevented reporting on sensitive military information, low morale among troops, and poor efficiency in the war effort (Bernhard, 1993). At the height of the Cold War, a number of network reporters simultaneously served as both news correspondents and CIA agents with the knowledge of both their broadcast and government employers (Bernhard, 1993). Such actions helped to achieve government goals that governments were not able to fulfill on their own. Perhaps not surprisingly, governments at the beginning of the twentieth century argued that communications should be seen predominantly as the work of governments, particularly by the military (Hanson, 2008).

Government controls, government censorship, self-censorship, and the loss of venues of the public sphere can all have a negative impact on public education and discourse, both within the broad public sphere and within small worlds. However, the corporate control of media—thereby "news" and many important channels of information flow—presents perhaps the largest challenge to the success of the public sphere (Herman & McChesney, 1997). In the early 1990s, a mere twenty-three corporations were in control of the vast majority of newspapers, magazines, television, radio, books, and movies (Solomon, 1993). The rise of the twenty-four-hour news channels—like CNN, Fox, and MSNBC—also presents challenges for the public sphere. The twenty-four-hour news cycle can breed cynicism about the political process in viewers, as little gaffes and minor issues can be elevated to major stories as a way to fill time and generate interest in the coverage to the detriment of the perceptions of the importance of the political process. While such stories—like the gossip and scuttlebutt that can circulate in small world settings via small talk and chitchat—may have information value in specific contexts and certain kinds of small world situations, in the lifeworld as a whole they often seem particularly egregious, actually limiting the amount of time and attention that might be paid to other issues of broader import and interest.

Even with the explosion of the Internet as a source of information, the number of major corporations controlling the movement of information throughout the lifeworld has decreased in the past fifteen years, as a result of international mergers and reductions of the few legal barriers to multinational media monopolies, resulting in such globe spanning—though not necessarily successful—communications behemoths as AOL Time Warner, News Corp, and Universal Vivendi. In terms of mass media outlets, "for all that the number of outlets has grown, the number of people engaged in collecting original information has not" (Project for Excellence in Journalism, 2005). As a result, information provided by the commercial media remains inextricably linked to issues of profitability to providers, advertisers, and marketers and to government interests rather than to the actual information needs of small worlds, larger information worlds, or the lifeworld as a whole (McChesney, 1993). And the increase in media choices also has not led to greater political engagement or awareness. Radio, television, and the Internet—both news and advertising—have not improved levels of voting; in the nineteenth century prior to these communication media, 80% to 90% of eligible voters actually voted in national elections (Smith, 2004). The wider range of entertainment choices has drastically reduced the viewership for—and therefore the awareness of—news programming (Prior, 2007). Fans of entertainment media devote more time to entertainment, excluding themselves by choice from politics, as opposed to times when newspapers excluded those who could not read. Ironically, this self-exclusion is occurring when political knowledge has never been easier to access (Prior, 2007).

While consumers' choice to focus on entertainment information certainly says something about perceptions of information value across many small worlds (as people valorize such information above political or social information), it is also a rather striking example of how information worlds' perceptions of information value can lead them to actively disregard information that may, in fact, have a demonstrable and important impact on their lives. The point in this discussion, as always, is not that information related to entertainment (or other "trivial" matters) is a bad thing or is a clear-and-present danger to the health of the lifeworld or other information worlds, but rather that all types of information co-exist across the range of information worlds, meaning different things to different people and playing a variety of roles in the information lives of individuals.

Interestingly, considering the degree to which it allows access to information far beyond what any government would sanction, including not only politically subversive materials but materials, like pornography, that may violate the social norms of many information worlds, the Internet also has stronger government ties than previous revolutions in ICTs. The private sector was mostly or totally responsible for the creation and adoption of the printing press, the telegraph, the telephone, radio, and television; in contrast, the Internet would not exist but for the U.S. military (Klotz, 2004). The influences of commercial and governmental interests even shape the education of future professional journalists. As one journalism professor noted nearly two decades ago, "We all know, whether we're candid enough to acknowledge it or not, that the advertising, news and public relations industries that provide employment for our students—plus other benefits—expect us to follow the 'company line' on issues involving the special interests of mass communication" (DeMott, 1990, p. 9).

MEDIA AND THE CONTEMPORARY PUBLIC
SPHERE IN THE UNITED STATES

The global impacts of the Internet and mass communications have been far-ranging and have had many divergent effects on the information worlds of various cultures and societies (Anderson, Brynin, Gershung, & Raban, 2007; Chakravartty & Zhao, 2008; Gough & Stables, 2008). However, all are linked by the fact that the increased reach of information technology and the consolidated power of commercial media are a potent combination in shaping the information available in the public sphere. In fact, "a little objective reflection should tell us that all societies are information societies, but also that they are information societies in different ways" (Duff, 2000, p. 172). In democratic societies, the relationship between media, information worlds, and the public sphere is particularly important. Such societies are defined by the periodic selection of representatives in most cases, meaning that "the public has to know what is going on and the options that they should weigh, debate, and act upon" if the system is going to be effective (Herman & McChesney, 1997, p. 3).

The role of media could not be more significant in providing information for small worlds and lifeworlds to consider, debate, and resolve meaningful social and political issues. However, the commercial media seem to have entered a period of enhanced efforts to shape discourse, with the media of the United States perhaps typifying the contemporary trend of news with an agenda. In many ways, it seems natural that commercial media would not attack the corporations that pay for advertising or the government that regulates it. For example, it is unusual for local news media to do hard-hitting critical examinations of the most powerful families and commercial institutions in their own communities (Nichols & McChesney, 2005). At the national level, consider the potential complexities raised by NBC—one of the major providers of news on television and the Internet—being owned by a General Electric (GE)—a corporation that does very large amounts of business with the U.S. government and many other national governments. These complex entanglements between corporate interests and the media were clearly demonstrated in 2009 when MSNBC signed a long-term, multi-million-dollar agreement with the coffee chain Starbucks to sponsor their morning news broadcast, which was rechristened "Morning Joe Brewed by Starbucks."

Further, to promote the appearance of objectivity, relevant political context—how political information is inextricably linked to specific information worlds and specific norms and values—is often avoided (Nichols & McChesney, 2005). The perspectives of the media have also been shaped by persistent critiques of the media as favoring the political left, encouraging a return to journalism serving the interests of the small handful of corporations that own the vast majority of media outlets. The concerted effort by the Republican Party to shift discourse through the creation of clearly right-wing commercial media and the common use of right-wing experts on other media proved enormously successful at making the perspectives of media more oriented toward the interests of the information world of the Republican Party (Nichols & McChesney, 2005). This increased politicization of news content has been greatly facilitated by a long-term shift in the format of news. Since the early 1980s, news programming has moved from a scripted narrative to more free form interactions (Clayman, 2004), allowing the norms and values of particular small worlds to be more easily inserted into ostensibly "objective" news content in the guise of presenting debate and dialogue, giving authority to those norms and values. As a result, while a cursory glance at such news programming might suggest the presence of an increasing diversity of voices, these changes have actually significantly constrained the representation of the full spectrum of information worlds in the media.

The media has also fallen victim to its close relationship to government. While meant to promote objectivity, professional journalism prioritizes official government sources and thereby focuses news on what the government wants. Commercial media corporations also want to avoid angering the government to perpetuate limited regulation of media corporations. This relationship constricts the information about government activities that the media feels it can provide. By creating limitations on access to information by the media, the government has strongly shaped what is discussed in the media and how it is discussed, reinforcing the goals and values of the information world of the government (Hiebert, 2003, 2005).

In spite of widespread beliefs to the contrary, in a democratic society, commercial media is not—nor can it be—objective; instead, it can heavily shape the information available in the public sphere and show how that information is contextualized and perceived. The leanings of current media, reflecting the information value of the media, are based on "what (and who) gets preserved—written down, printed up, recorded, filmed, taped, or scanned—and why" (Gitelman, 2006, p. 26). These practices, through which information value is translated into information behavior, are key instigators of international events, shaping how information is considered and the actions taken based on it. In other words, how the information world of the media packages and sells its own perceptions of information value and appropriate information behavior not only can ripple throughout the lifeworld but also can impact and shape small worlds' capacity to gather, use, and share information.

The ways in which the media handles information can have the gravest of consequences across the information worlds of democratic societies. Since the Civil War, the major wars that the United States has engaged in have been driven initially by executive interest in war and public resistance, but in most cases, "the White House ran a propaganda campaign to generate public support for going to war" (Nichols & McChesney, 2005, p. 41). These campaigns are designed to encourage emotive responses rather than encourage public debate about conflict. "Americans have always been reluctant to mobilize for military action. When they do support military action, they typically have little understanding of the international scene, and they do not follow their leaders into war unless the justification is in the name of freedom" (Brenkman, 2007, p. 46).

In times of war, James Madison and many others after assumed that the media would serve as the most significant check on the power of the executive. The contemporary commercial media, however, may not be capable of performing this vital function; when the importance of being visibly patriotic is a strongly promoted—and even unquestionable—information value and social norm, for the media to exercise its role as a check on government is not without significant risks, not least in ratings wars. This compliance is in part derived from the coziness of commercial media and government, but more importantly it results from critiques of left-wing media bias. "An incessant aspect of the right-wing critique of the 'liberal' news media is that journalists are insufficiently patriotic; this translates into journalists being extrasensitive to prove their nationalist credentials" (Nichols & McChesney, 2005, p. 49). In this scenario, the political leanings of the media can become a much stronger information value than the accuracy or depth of their reporting.

In contrast to the perceptions of rampant left-wing bias, however, a large and vocal segment of the media has held distinctly conservative perspectives since the dawn of radio. The far right of the political spectrum has always had a strong presence in American broadcasting, with far right broadcasters and station owners—such as Haraldson Lafayette Hunt and G. A. "Dick" Roberts—using radio as early as the 1920s and 1930s to spread their messages (Chester, 1969). Roberts instructed the reporters and commentators at his station to always place many prominent democratic and progressive figures in an unfavorable light, preferably by linking them to Judaism or Communism, while simultaneously supporting the actions of the Ku Klux Klan; Hunt pioneered the openly one-sided news broadcast, developing programs for his stations that promoted the social norms and political opinions of the far religious right (Chester, 1969). "Throughout American history the censorship of left-wing philosophies and concepts has been far more common than that of right-wing ones" (Chester, 1969, p. 218). During World War II, for example, the two groups who were the focus of a government ban on discussing controversial public issues were labor organizations and cooperatives (Chester, 1969).

Following these historical trends of a prominent far right perspective in the media, carried on today by conservative talk radio and Fox News, the majority of the media made no attempts to do any investigative reporting related to the buildup to the invasion of Iraq. In March 2003, at the launch of the preemptive Iraq War, President Bush held a press conference where the elite of the national media made no inquiries into the questionable legality of preemptive war, the obviously flawed case for war, the lack of a plan for an exit strategy, or the costs of this military endeavor. As such, the information world of the media chose to preserve their veneer of patriotism rather than act as a check on the government by ensuring that adequate information was available for citizens to debate the validity of the war.

The media even helped to promote the Iraq War to serve its own mainstream purposes. "After September 11, political leaders and the media cultivated and relentlessly fertilized an all-pervasive fear within the American body politic" (Brenkman, 2007, p. 8). Many networks also used the buildup to the war in Iraq and the strange "liberation" itself as a moneymaking venture—using focus-group-tested titles like "Countdown to War" and "Showdown with Iraq" and speaking of the war in entertainment metaphors to significantly increase viewership or readership (Grusin, 2004). Networks ordered uplifting and feel-good music for the newscasts related to the invasion of Iraq, with CBS and Fox using rhythm tracks intended to provoke feelings of excitement (Smith, 2004). Overall, cable news has greatly benefited from terrorism coverage, with ratings going up at least 25% when terrorism is at the front of the news cycle (Sunstein, 2005). The overhyping of threats, such as the widespread exaggerations of the level of danger presented by anthrax in 2001, can be seen as bad reporting, but also as a means to increase ratings (Murray, Schwartz, & Lichter, 2002). Again, in terms of information value, such practices foreground the packaging and presentation of stories instead of their actual information content.

Premediation of the Iraq War—the Bush administration constantly talking about the possibility of the war and causing the media to discuss the possibility—ensured that the public felt threatened and opted to return legislative power to the Republicans in the 2002 election (Grusin, 2004). This same process allowed the networks to increase their ratings by utilizing public fear as a positive information value in shows about the possibility of war, and to market-test how best to present the actual war to audiences once it started to earn the largest audience share. Bush's doctrine of preemptive war required a preemptive media—playing the role of the compliant information world rather than that of the independent media—who went along with Bush by presenting war as an inevitable aspect of the future (Grusin, 2004). And the media obligingly presented the future prospect of war as if it was already a present reality. The media embedded in Iraq was the largest in the history of the military, with over 500 news personnel joining with the military. Many media outlets used this embedding to reinforce the pending war as entertainment through metaphors about front row seats and the use of videogame-style graphics.

Some media outlets even promoted the war for ideological reasons and to establish certain beliefs and values as normative across small worlds. Rupert Murdoch spoke out as a staunch advocate of the war, and soon after all 175 newspapers in his media empire began running pro-war editorials and news coverage (Greenslade, 2003). Similarly, while most Clear Channel stations lack their own news departments, the company did manage to sponsor eighteen pro-war rallies before the invasion of Iraq. Clear Channel is owned by the man who bought the Texas Rangers from George Bush (Smith, 2004).

This tacit and explicit support of the war by the media was part of a generally very successful strategy of controlling the information world of the media by the Bush administration. "Bush entered the White House with more experience dealing with the press than any previous president. What's more, he has used this experience to manage press relations through a combination of personal charm (and sometimes pressure), message discipline, and the rigid control of both press access to administration sources and control of leaks coming from sources within his administration" (Mueller, 2006, p. xv). Drawing from his experiences as a governor and the owner of a professional sports team, Bush controlled the press through affability, deference in the post-9/11 environment, exploitation of reporters' tendency to avoid evaluating the truth of political statements, the dependence of reporters on the White House for news, political pressure on dissenting journalists, and shallowness of mainstream news coverage (Fritz, Keefer, & Nyhan, 2004). To help control the message, the events around Bush tended to be completely scripted, from cabinet meetings to visits to town halls to policy forums (Suskind, 2004).

MEDIA AND THE CONTEMPORARY PUBLIC SPHERE AROUND THE WORLD

Many of the same trends of media monopolies, media consolidation, and media conglomerates with a specific perspective conveyed through their reporting seen in the Untied States are, in fact, international trends, and the impact of these trends on information worlds at all levels is a global phenomenon. Many look to the BBC and similar government-funded media as responsible media entities that actually serve the public sphere, but corporate controls, government controls, and self-censorship also affect these media organizations. While different philosophies of national governance have resulted in many different contexts for media, the move toward multinational media corporations has served to flatten the differences between the information worlds of the media, particularly in industrialized nations.

In a global society interconnected by ICTs, perceptions of international incidents can be dramatically affected by media manipulation, even if only one media corporation is engaged in such manipulation. In the war between Israel and Hezbollah in 2006, bloggers discovered that Reuters was providing newspapers and websites around the world with digitally doctored photos—known as fauxtography—that were designed to greatly exaggerate the damage in Lebanon caused by the Israeli military, such as changing a photo with one column of smoke to have three columns of smoke (Usher, 2008). Though the tenacity of the bloggers eventually caused Reuters to admit to these activities, the photos had already had a large impact on international public opinion about the conflict. Thus, even though this story demonstrates the ability of small world information outlets to sometimes have an impact on the dominant information world of the mass media, it also offers a more cautionary note: even though the bloggers ultimately achieved their goal of fact-checking the information offered by Reuters, the damage had already been done, and the initial misinformation ultimately carried more value across the lifeworld than the retraction.

Though international trends in media can be viewed from a range of perspectives, the most compelling way to understand their roles is to focus on the political economy of media organizations—the level of government control of the media, the involvement of the media in the political process, the partisanship of media, and the corporate motivations of the media (Chakravartty & Zhao, 2008; Shim, 2008). While it is clearly impossible to catalogue the media environment in every nation in this context, it is instructive to examine the international trends building from more democratic nations to more authoritarian nations to see the range of ways media can affect the public sphere.

The British government created the publicly funded, non-commercial monopoly for the BBC in the 1920s, mandating that the BBC act as a public service relatively free from government control—that it, in other words, function as a semi-independent small world. By the 1950s, ITV provided the first commercial competition for the BBC, though ITV was still heavily regulated by the government. However, the conservative Thatcherite movement of the 1980s worked to erode the monopoly and quality of the BBC, reflecting an information value of "strong government and a subservient media" (Herman & McChesney, 1997, p. 167). These actions to reduce the power of the BBC were concurrent with efforts by conservatives in the United States to diminish the power of PBS, using many of the same tactics—large budget cuts coupled with highly partisan appointments to supervisory and oversight positions of persons opposed to the norms and values of the organization they were overseeing (Freedman, 2008; Herman & McChesney, 1997). The introduction in 1989 of satellite television, with no mandate to provide anything other than popular entertainment, further eroded the competitiveness of the BBC, which now has its own commercial and pay channels to keep itself financially viable. The BBC, still funded partially by taxes, receives a guaranteed income of approximately 3 billion pounds annually, an average of 116 pounds per citizen (Aitken, 2007). And, despite the competition, 93% of citizens of the United Kingdom use at least one BBC service each week (Aitken, 2007), making its reach into small worlds across the country significant.

Conservative media barons Rupert Murdoch of Australia and Conrad Black of Canada each created powerful information worlds in the form of media monopolies through control of many newspapers, television channels, magazines, and Internet outlets in the United Kingdom (Freedman,

2008). These media empires were made possible by loosening of the restrictions in the United Kingdom—as in other nations such as the United States and Australia—on the number of media outlets that can be owned in a particular market by one corporation; these rules were originally based on the reasoning that supporting multiple small worlds as reflected in the diversity of new media channels reduces the threat of too few perspectives being presented through broadcast media (Hitchens, 2006).

With staunchly conservative views and strong ties to the conservative Tory party, many commercial media outlets became de facto Tory propaganda organs (Herman & McChesney, 1997), thus serving the agenda of a very particular information world with a very particular understanding of information value and a very specifically defined set of social norms. After the Tories' major electoral success in 1992, their treasurer commented that "The real heroes of the campaign were the editors of the Tory press. . . . This was how the election was won" (quoted in Herman & McChesney, 1997, p. 169). When Tony Blair, the former Labour Party Prime Minister, first ran for the office in 1995, he traveled to Australia to meet with Murdoch to try to persuade him to have his media outlets be less overtly supportive of the Tory party in exchange for dropping increased government control of commercial media from the Labour platform (Freedman, 2008). On the other hand, the BBC content has also been criticized for bias, but bias that favors the norms and information value of the Labour Party, liberal positions, and the European Union (Aitken, 2007).

Other nations have created BBC-like entities in an attempt to preserve a media information world free from corporate control—though not free from government control—only to see them also overrun by commercial media. Canada's CBC is an example of this situation, with the Canadian media market dominated by multinational corporations and the CBC now affiliated with a range of commercial stations. The news-gathering and -reporting functions of the CBC and commercial channels in Canada have directly paralleled those in the United States, resulting in all the same problems (Herman & McChesney, 1997). Further, the market influence of Murdoch and Black has helped to swing general media discourse to the right, as in the United States and the United Kingdom, though Black's influence has likely dwindled since 2007 when he was sent to prison on charges of mail fraud and obstruction of justice.

At least none of these media barons, however, have ever been elected to lead a country, as has occurred in Italy, which has elected its leading conservative media baron as prime minister multiple times. Italy has a long tradition of corporations controlling major media outlets, with many industries also owning media outlets, creating a strong information value of active support of business interests in the media. Silvio Berlusconi's companies not only control about 50% of the television audience share, they also control the largest movie production and distribution company, movie theater chain, book and magazine publisher, advertising chain, newspaper, and

TV weekly, as well as many other newspapers, a department store chain, a major investment firm, and a prominent soccer team. During his first term as prime minister—when, that is, he played the dual role of figurehead and leader of two particularly powerful and influential information worlds—he disassembled Italian public television, his only real media competitor. All of these outlets and types of media control likely facilitate Berlusconi's continuing electoral successes and his efforts to consolidate the Italian media into a single information world with uncontested control of the flow of information through the country.

Media in most industrialized societies is dominated by the information values of commercialization and politicization, with extreme impacts on the goal of objective information dissemination required to sustain the health of the public sphere. In many nations, public broadcasting services have been forced to recast themselves as public service media, working to incorporate diverse media and technologies to reach audiences in new ways to better compete with commercial outlets (Bardoel & Lowe, 2007; Freedman, 2008). Even supra-national organizations are working toward these ends. The policies of the European Union (EU) have served to greatly reduce public broadcasting in all its member nations and promote commercial broadcasting (Sarikakis, 2008). More distressingly, EU policies have also worked to force media to act as a channel to promote and legitimize the EU governance structure as a centralized information world through requirements for use of EU symbols, promotion of EU identity, and delivery of EU messages (Sarikakis, 2008).

Other approaches by governments have been to drastically limit the number of media outlets to ensure the media functions as the government prefers. For example, between 1968 and 1993, Israel allowed only one television station as a policy decision to control content and promote educational goals and state interest (Katz, 1996; Weimann, 1996). The outcome of all of these efforts has been a broad-based focus on the media as an outlet serving the needs and values of a single, extremely powerful, information world while, in the process, imposing severe limitations on the capacity of smaller, more localized, information worlds to maintain their own identities and to have their voices heard across the lifeworld.

Under authoritarian governments, the situation is worse, not surprisingly. The media are often controlled by the government to function as a part of the government, using media channels and outlets as a means both of control over information and of social control (Hanson, 2008). Such consolidation of government and the media establishes an information value of tight control as a mandate across all information worlds in their countries and actively—and sometimes completely—constrains the ability of small worlds to maintain any kind of independent presence or voice. The Russian media has worked, since the initial election of Vladimir Putin as prime minister, to promote nationalism and shield the populace "from [the] nonofficial viewpoints" of small worlds (Koltsova,

2008, p. 68). In Hugo Chavez's Venezuela, the media is regulated under a policy that includes legal penalties for the criticism of politicians (Duffy & Everton, 2008). More than three dozen nations filter Internet access within their borders to control what information citizens can access and exchange (Zittrain & Palfrey, 2008). However, the most thorough use of media to limit the capabilities of small worlds and to control citizens may be the Chinese government. The heavy hand of the Chinese government controls information flows and dictates the message provided by the media through "state repression and disciplinary technologies," exemplified by the Great Firewall of China that radically limits both the news and political, environmental, and social information that Chinese citizens can access online and the ability of those citizens to post their own materials and add their own voices to the nation's information world (Zhao, 2008, p. 25). In 1999, China began arresting citizens for writings they posted on the Internet; in 2002, China closed all but 200 of the 2,400 Internet cafes in the country (Klotz, 2004).

MEDIA, INFORMATION RIGHTS, AND INFORMATION WORLDS

Constraints on the free spread of information imposed by corporate media run contrary to a line of legal reasoning that has developed in a number of nations. In the United States, the Supreme Court has explicitly stated that "the Constitution protects the right to receive information and ideas" (*Stanley v. Georgia*, 1969, p. 564). Further, in *Lamont v. Postmaster General* (1965), the Supreme Court held unconstitutional a statute that required the U.S. Postal Service to prevent delivery of any communist materials unless the recipient of the materials had affirmatively requested that such information be delivered, holding that having to actively request the materials, which would otherwise have been freely available except for the statute, was an unconstitutional burden on free speech. Without this protected right to access and receive information—even if that information reflects the possibly "dangerous" positions of disparaged small worlds—the overall right to free speech could be significantly diminished (Mart, 2003).

These types of arguments are not unique to the United States. Governments and public interest organizations in a number of different countries have articulated some version of an explicit information value of maintaining information rights—a legally definable set of rights to expression, access, and control of information (Caidi & Ross, 2005; Ross & Caidi, 2005). This argument becomes even more salient if one agrees with the assertion that the right to communicate on the Internet should be framed as a human rights issue (McIver, Birdsall, & Rasmussen, 2003). Thus, even as the positive information value articulating a set of information rights for the individual has become more expansive, the channels of access through corporate media have become more constricted.

Unfortunately, consumers of information often adapt their information behavior to meet the options that ICTs provide, "understanding and normalizing" the levels of access the ICTs allow (Lastra, 2000, pp. 6–7). In this sense, the deployment of ICTs through a lifeworld can exert normative influences, as consumers begin to see the limitations built into such deployment simply as "the way things are." As a result, the prevailing information world of the mass media can have a direct and limiting impact on the range of normative behaviors and perceptions of information value both within and across small worlds; constrictions of information flowing through the media act to shape the amount and types of information that citizens will seek and expect to receive in their small worlds and across the lifeworld. As fewer citizens actively search for and exchange information about meaningful issues, it becomes a threat to the health of the discourse in the public sphere.

Gilman-Opalsky (2008) casts this as a choice between the efforts of isolated individuals and mass cultural movements: "arguments challenging the generally accepted meaning need to take flight and take hold on a mass scale. This step cannot be bypassed by the intention or action of an isolated individual; only collective action in the public sphere can deconstruct the prevailing signification" (p. 339). However, an analysis rooted in the theory of information worlds suggests a different kind of double-sided situation, apart from the binary of individuals versus mass action: media hegemony (and the resultant constrictions of information exchange) threatens not only the efficacy and resources of individuals and the culture at large but also the health of information worlds at all levels. Conversely, the health of the public sphere and the broad lifeworld are dependent not on either individuals or the mass, but on the health and potential of many worlds at both the micro and meso levels. Effective actions within such worlds, as participants in those worlds insist on making their voices heard, may have a powerful salutary impact on the macro level of both the public sphere and the lifeworld.

In a practical sense, loss of information resulting from the consolidation and the commercialization of media around the world may be best demonstrated by the impacts on the small worlds of local communities. As traditional local media have disappeared (e.g., local newspapers) or been purchased by multinational corporations (e.g., local radio stations), the number of sources of information in most communities has plummeted (Klinenberg, 2007; Vaidhyanathan, 2006). For example, when a train derailment sent a cloud of poisonous gas floating toward Minot, Minnesota, local officials were unable to use media to alert the townspeople because all six local radio stations were owned by Clear Channel Communications, and the stations played only feeds of preprogrammed music from a distant source. Local officials were unable to get any response from Clear Channel's national offices to announce the impending danger; as a result, one resident died from exposure to the gas and more than one thousand were injured (Klinenberg, 2007). This is one example of a number of ways that the consolidation of media has created barriers to communication and information dissemination in emergencies (Jaeger, Fleischmann, Preece, Shneiderman, Wu, & Qu, 2007; Jaeger, Shneiderman, Fleischmann, Preece, Qu, & Wu, 2007).

However, the limitations on information flows through the media to the public sphere create much broader threats to the lifeworld as well as to the functioning of a democratic society as a whole. Media policy encompasses issues of maintaining freedom of expression, structuring regulatory bodies, managing the spectrum, ensuring broadcast diversity, framing and analyzing issues of social consequence, engaging informed political debate, and monitoring government inaction and corruption (Buckley, Duer, Mendel, & Siochru, 2008). And, the media frequently base their decisions on business calculations rather than social ones. For example, when faced with concurrent requests for user information from the U.S. government and censorship requirements from the Chinese government, Google challenged the U.S. government and complied wholly with the Chinese government (Fry, 2006). While Google was not going to be shut down by the U.S. government for protecting the privacy of its users, Google could only have a presence in China if it was willing to sacrifice the information access capacities of users to the satisfaction of the repressive Chinese government. And Google is not alone in their compliance with the censorship demands of the Chinese government, as other companies such as Yahoo! and Microsoft are willing to meet demands of the Chinese government; Yahoo! even gave to the Chinese government in 2003 information that was used to imprison one of their users (Dann & Haddow, 2008). Indeed, while the writing of this book was in the final stages, yet another controversy erupted over Chinese demands that access to information of various sorts be constrained by requiring that technological limits on access—via software known as "Green Dam"—be immediately built into computers sold in China (Jacobs, 2009b).

Without sufficient information about important political and social issues, the ability of citizens to deliberate and discuss throughout information worlds at all levels is greatly impaired. Democracy can only exist when citizens actively discuss issues of governance (Dewey, 1959). Without "reasoned public discussion of political questions," democracy cannot succeed (Rawls, 1996, p. 224). Without consistent exposure to multiple viewpoints about a political issue, people tend to have a limited understanding of both the political issue itself and the broader policy processes surrounding it (Hofstetter, Barker, Smith, Zari, & Ingrassia, 1999; Jaeger, 2007; Prior, 2007). Yet, technology has made possible—and media corporations have profited from—the ability of individuals to narrowcast the information they are exposed to on radio, television, the Internet, and even government information to the extent that it is possible to avoid information that does not fit with the social norms and information value of the small world in which the individual exists.

In short, democracies "rest on the assumption that citizens can govern themselves because they are informed" (Hacker, 1996, p. 213). The Supreme Court has noted that there is "practically universal agreement" that the primary intent of the First Amendment is "to protect the free discussion of governmental affairs" (*Landmark Communications, Inc. v. Virginia*, 1978, p. 838). However, without sufficient sunlight being shed on political processes by the media, it is very difficult for citizens to have adequate information about social and political issues to deliberate and make decisions at any level. Insufficient information about important issues can reduce deliberation in small worlds and in the lifeworld, constricting discourse across the information worlds in a society.

Media, in the traditional sense, are now mostly controlled by a few specific and extremely influential information worlds, and thus serve as a powerful tool for imposing their own social norms, social types, information behavior, and information value across the lifeworld. As widely trusted arbiters of information, the ways in which the news media treat information can have enormous impacts throughout information worlds, large and small. By influencing the perceptions about certain information in many small worlds, the media can shift the overall social perception of or value given to that information across information worlds.

When the media is performing its intended function in a democratic society, it increases exposure for the perspectives of many divergent small worlds within the lifeworld. However, when it does not meet this function, it can work as an external influence over many small worlds, actively limiting the ability of those small worlds to access and use information. As was demonstrated with the Iraq War, the combination of open support for the war by some of the dominant worlds within the media and the desire to not look weak by questioning the war in many others, ultimately flooded the lifeworld with a particular vision of information value, promoting widespread initial enthusiasm for the war in a great many small worlds and in the lifeworld of the United States as a whole.

It has been suggested that freedom of expression stands as "a sort of 'moral Alamo'" (Giacomello, 2005, p. 106). In a society that has given away many rights for various reasons, freedom of expression may serve as the one bulwark of democracy that the citizenry would never relinquish. However, if the media fails to serve its appropriate function, it becomes much harder for deliberation and discourse about important political and social topics to flourish across the information worlds of that society. The next chapter explores how an information world of those with political power can reshape entire other worlds, both small and large, by using politics and policy to impose the social norms, social types, information behavior, and information value of that one world across the lifeworld. However, the chapter also explores how small information worlds with very limited power can creatively use ICTs to try to try to shift the perceptions of other small worlds and the lifeworld alike.

8 Information Worlds and the Political World

The rules established by the government in relation to information are the most significant drivers of the amount of information available to a society and the ways in which that information can be accessed and exchanged, whether across the full lifeworld, or within or between individual small worlds. While governments have always worked to influence discourse in accordance with their own norms, values, goals, and benefits, ICTs have greatly increased the potential to use information for large-scale political goals. These efforts can start at the lifeworld level or at the small world level in attempts to shape discourse across information worlds. Building on the historical uses of information for purely political purposes ("information politics"), this chapter compares the international efforts of the George W. Bush administration with the highly localized efforts of the Zapatismo movement.

As a particular kind of information world, a government that engages in information politics is truly working to make the uses of information in society conform to their own vision of information value, which serves to significantly alter the meaning of information access and exchange. Policies that limit information access serve multiple purposes, accomplishing both policy goals and partisan political goals. After the terror attacks of 9/11, the Bush administration employed public policy and political pressure to implement a series of major changes to the amount of information available to the public, the ways in which the public can access and exchange information, as well as the amount of information that the government can collect about individuals. While the Bush administration demonstrated a strong information value of placing tight controls on information prior to 9/11, the "war on terror" was used by that administration as a means of reshaping information in society. The Bush administration, then, presents a case study in how a government can try to use policy, politics, censorship, intimidation, the legal system, and new technologies to influence how information moves among information worlds (large and small) in a democratic society.

The perspective of the Bush administration appears to have been based on an information value that posits that access to information, in general, should be very limited to all but a small number of members of a very specific small world and tightly controlled in a centralized manner by the executive branch of the federal government. It also included the belief that information that is made available for access should fit with the administration's own beliefs and values. Taken together, these policies have had significant impacts on the amount and types of information available for access as well as presenting serious questions about the long-term impact on democracy in the United States and internationally. In hindsight, the information policies of the Bush administration cannot be separated from their practical political goals, such as limiting public discussion of administration actions, protecting private interests, and increasing public fear. All of these uses of information value and their attendant impact on information behavior across the country challenge the core ideals of democracy, a concept of government that is based on the premise of an informed citizenry. Further, quite a few other democratic governments around the world have followed the lead of the Bush administration on many of these issues.

In contrast to the efforts of the Bush administration are the efforts of those employing social networking and websites to exchange information, using the Internet as a means to impact discourse in the public sphere. The Zapatista movement, for example, has used postings on the Internet to try to encourage reformation and revolution in Central America for years, creating a public sphere out of a reading public—a true war of words (Gilman-Opalsky, 2008). These two very different approaches demonstrate attempts to shift political discourse in information worlds in favor of the political organization through influencing either the lifeworld or small worlds.

THE HISTORY OF INFORMATION POLITICS

Governments generally try to influence peoples' perceptions of information value through the information they choose to release and emphasize, and by how they release it (Sunstein, 2005). However, the levels of control they try to exert and the degrees of success have varied by types of government and historical epochs. From the Middle Ages through the 1950s, few questioned the notion of the nation-state or the justifications for nationhood—it was simply accepted as the way of the world, or, to use the terminology of the theory of normative behavior that has been adopted for the theory of information worlds, it was a nearly universally accepted social norm (Toulmin, 1990). The idea of the nation is at the foundation of any understanding of the modern world (Chernilo, 2007). A defining element of nation-states is a central government that guides the nation and its society, and throughout much of history, these governments have worked to ensure that they stay in power.

In the centuries before widespread literacy or mass printing, control was generally localized, with local leaders dominating a small area and reporting up to the higher levels of central control. The rise of the printing press and mass-produced books, however, radically changed how central governments had to work to control the availability and movement of information in society. Moveable type and the subsequent mass production of printed books, thus, not only spread reading but also served to establish the concept of the centralized nation state. Books were printed in the most commonly used languages, rather than the hundreds of dialects and minor languages at use in localized areas of Europe. As a result, printing encouraged the formation of cultures and nations around major languages, simultaneously and ironically fostering both communication and barriers to communication.

Printed documents encouraged more structure and centralized control, a single point of view to be expressed by the government, and a single language of law and governance, resulting in a greater national orientation by citizens and a move away from feudal connections (Hanson, 2008). "This shared sense of group identity gradually evolved into the ideology of nationalism" (Hanson, 2008, p. 16). The new paradigm precipitated by the development of printing also clearly gave governments an incentive to try to control information: it has, in fact, been noted that the invention of the printing press led directly to the invention of government censorship (Pool, 1990). It also led to governments taking steps to institutionalize a variety of means to control information both in terms of their subjects and in terms of competing governments—collection and storage of information, supervision of activities of other people, and application of information gathering activities to monitor and ensure compliance of people under surveillance (Dandeker, 1990; Giddens, 1985; Howard, 1983).

The innovations of democratic governance, however, interceded in these efforts of control of information by positing an information value rooted in the idea that citizens' rights of access to information were essential. "Political theory has given a twofold answer to the question of [state] legitimacy: popular sovereignty and human rights . . . ground an inherently legitimate rule of law" (Habermas, 2001, pp. 115–116). For democracies, "the co-originality of liberty rights and the rights of citizens is essential" (Habermas, 2001, p. 117). Guarantees that originated in the United Kingdom's Magna Carta and the United States Constitution and Bill of Rights have spread throughout democracies and republics around the world. The rights to freedom of expression, press, assembly, access, and other rights like privacy and open government have created protections of access to and exchange of information in many societies. Without access to the right information it is possible to have free expression yet not have a democratic government (Berlin, 1996). As such, the key underlying concept of these democratic rights is necessity of access to meaningful social and political information (Jaeger, 2005, 2007; Jaeger & Burnett, 2005).

More recent ICTs, however, have greatly increased the ability of governments to control information. By improving lines of communication,

ICTs—like the telegraph and then telephones—were able to provide a tool of increased effectiveness in colonial administration and control, enhancing "the power of the rulers over the ruled" (Hanson, 2008, p. 19). The Internet has been a particular boon for government control of information in some parts of the world, perhaps because the Internet's ability to support the information access and sharing of many small worlds makes it a particularly attractive target to those who would control information, both across the lifeworld and within the boundaries of specific small worlds.

More than three dozen nations filter access to the Internet, primarily concentrated in East Asia, North Africa, the Middle East, and central Africa (Zittrain & Palfrey, 2008). Malaysia and Saudi Arabia began censoring Internet access in their countries as official government policy in 1999, with Saudi Arabia announcing the implementation of a strategy to screen, monitor, and censor the Internet usage within the country through the King Abdulaziz City for Science & Technology (KACST); the same year, China began arresting citizens for writings they posted online (Klotz, 2004). In 2002, Turkey passed a law against "airing pessimism" online, while China closed all but 200 of the 2,400 Internet cafes in the country (Klotz, 2004). In the United States, public libraries and schools receiving federal funding have had to filter their Internet access since 2001, creating disparities in levels of access available between school and libraries that need government funds and the wealthier schools and libraries that can forgo such funds and their related filtering requirements (Jaeger & Yan, 2009).

These nations filter for a range of social, political, and security reasons, blocking materials such as information related to free expression, health, human rights, economic development, environmental issues, religious beliefs, and other nations, among many others (Zittrain & Palfrey, 2008). Further, the advent of government websites has allowed many nations to use these websites as a tool of propaganda to influence both their own people and the perceptions of other nations (Chadwick, 2001; Jaeger, 2005).

THE INFORMATION WORLD OF THE BUSH ADMINISTRATION

The importance of access to information for information worlds can be particularly acute when it involves political information, as is made clear by the policies implemented by the Bush administration, including the reclassification of previously unclassified information, the removal of information from federal websites, and the actions taken to influence scientific research. Such changes may best be understood in terms of information worlds, as they are particularly significant instances of information coming into conflict with the social norms and information value of a distinct information world—a relatively small world in terms of numbers of members, albeit, in this case, one with an especially significant level of power over information access for many other worlds.

The executive branch of the federal government is directed by a relatively small number of individuals—the president and his staff, the vice president and his staff, cabinet members, policy advisors, directors of executive branch departments and agencies, and their associated staff. None of these officials, other than the president and vice president, are elected—they are all political appointees, selected primarily because their views match those of the president. This relatively small, but incredibly influential, group of people comprises a small world with its own social norms, sense of information value, social types, and information behavior. The views of the members of this small world predominate the policies generated and the actions taken by the entirety of the federal executive branch agencies, which employ much of the federal government workforce and implement most of the policies of the federal government. While this small world in any presidential administration could serve as an important case for the study of social access to information, the aggressive information access policies of the Bush administration make it particularly apt for this discussion.

The social norms of the Bush administration were quickly apparent after it took office, as they focused on establishing clear boundaries between those who had access to information and those who did not. The social norms of the administration were such that they tried to keep most information related to their activities away from anyone not in their small world, frequently ignoring requests for information made under the Freedom of Information Act, and oftentimes not even acknowledging the requests (Committee on Government Reform, 2004). In the first few months of the first term of the Bush administration, the vice president became embroiled in a fight with the Government Accountability Office (GAO)—a legislative branch agency serving as the government's internal watchdog—over access to records from a hearing held by an energy task force (Relyea & Halchin, 2003). The executive branch went to federal court to fight against the GAO's request for information about the hearing.

Other information requests by the GAO or Congress itself that were rejected by the administration included information about communications between the vice president and the Department of Defense about contracts to Halliburton, documents about prisoner abuse in Iraq, cost estimates for the Medicare prescription drug plan, air pollution data, presidential advisor Karl Rove's meetings with executives of companies in which he owned stock, and information requested by the Congressional Committee investigating the 9/11 attacks (Committee on Government Reform, 2004). In the case of the Medicare information, the actuary of the Department of Health and Human Services was told by the administration that he would be fired for giving the requested information to Congress (Committee on Government Reform, 2004). This pattern quickly extended into the administration's public comments, as members of the cabinet and advisors began to take the unusual step of routinely refusing to testify before congressional committees (Relyea & Halchin, 2003), thus making it clear that the boundaries of their small world did not include governmental representatives outside of the close circle surrounding the president. Administration officials were given specific talking points about issues from which they were ordered not to deviate when discussing policies outside the small world (Suskind, 2004). When members of the cabinet, such as Treasury Secretary Paul O'Neill, did not closely follow the assigned script, they were forced out of their jobs (Suskind, 2004).

The information value embraced by the Bush administration was similarly narrowly defined—information related to national security or intelligence had a higher value relative to all other forms of information and was carefully controlled. The Bush administration extended the classification of documents, gave the authority to classify documents to many more executive agency directors, encouraged reclassification and retroactive classification of unclassified documents, lengthened classification periods for up to twenty-five extra years, and created a presumption of secrecy with government information (Barker, 2005; Feinberg, 2004). These policies were accomplished through executive orders and through far-reaching legislation, such as the USA PATRIOT Act and the Homeland Security Act, which the administration heavily lobbied Congress to pass (Jaeger & Burnett, 2005). The Bush administration also used regulations to make specific types of information unavailable, such as information from commercial satellites and vehicle safety information (Committee on Government Reform, 2004).

The orders and memoranda by members of President Bush's small world were particularly revealing about their belief in the value of tightly constraining information access. An October 2001 order from the Department of Justice explicitly instructed federal agencies to release as little information as possible and assured federal agencies that the Department of Justice would defend them from legal action whenever they withheld information (Office of the Attorney General, 2001). A March 2002 memo from the White House instructed agencies to withhold any information that might be sensitive but which could not fit the legal definition of information that could be classified (White House Office, 2002). In May 2003, President Bush issued an executive order that limited access to information from current or previous administrations, postponed the automatic declassification of documents, created a protection from release for any government information related to a foreign power, encouraged extensive use of reclassification of publicly available information, and eliminated the presumption of disclosure for requests for government information (White House Office, 2003).

The information world of the administration, in addition, explicitly implemented the practice of social typing in its collection of information about citizens in order to classify them. The laws that facilitated the administration's limitations on the release of government information have simultaneously greatly increased the ability of executive agencies to gather information about citizens and resident aliens (Jaeger, Bertot, & McClure, 2003; Jaeger & Burnett, 2003, 2005; Jaeger, McClure, Bertot, & Snead, 2004). Further, the federal government turned to external data aggregators to gather more information about citizens (Jaeger, 2007). Federal law bars government employees from creating databases like those of data aggregators, but government employees frequently searched such databases as part of their jobs, and many government agencies had contracts for access to the databases of commercial data brokers (Roberts, 2006). As a result, social typing activities became part and parcel of federal information policy.

Not surprisingly, these beliefs led to information behavior that strongly limited access to information by members of the public—as well as by others who were not trusted members of the administrative small world. The most striking example of such information behavior could be found in the Bush administration's significant efforts to assert control over scientific and research information, including preventing the publication of research findings, centralizing peer review of applications for government funding, pushing for self-censorship among scientists and academic publishers, requiring all funded researchers to get agency approval to publish findings from any unclassified military-funded research, threatening to pull funding from academic researchers whose findings did not meet their goals, and preventing scholars and researchers from a number of countries from entering the United States (Jaeger & Burnett, 2003; Jerome, 2002; Knezo, 2003; Simoncelli & Stanley, 2005).

The Bush administration also sought to fill scientific advisory committees with persons who shared their own perceptions of information value and social norms (Committee on Government Reform, 2004; Simoncelli & Stanley, 2005). The Bush administration tried to influence scientific studies by government agencies to conform to its own beliefs, by modifying EPA reports on global warming and climate change, ordering the CDC to remove information about condom effectiveness rates from its website and replace it with a listing of condom failure rates and the effectiveness of abstinence, instructing Department of the Interior scientists to disregard alternatives to administration policies related to mining, and preventing the Food and Drug Administration from approving the over-the-counter sale of an emergency contraceptive (Simoncelli & Stanley, 2005). Simultaneously, a number of scientists, when being evaluated for non-partisan government posts, were questioned about their political beliefs and whether they voted for Bush (Simoncelli & Stanley, 2005).

The attitudes of the small but potent information world of the Bush administration also reveal specific intentions about social access. First, the understanding of information value of this world was based on a belief that social access to information should, in general, be very limited. This value was at the heart of many of the policies designed to limit information access for all other members of society. Second, the information value of this small world also dictated the information behavior that resulted from the social access to information by members of the small world itself. Such information behavior was displayed in the attempts to influence scientific and other research studies and reports. By setting new parameters for scientific research and by filling scientific committees with people who share the information value of the small world, the administration tried to make research fit its own social norms.

THE INFORMATION WORLD OF THE
FOURTH BRANCH OF GOVERNMENT

Perhaps no single incident better defines the administration's information value of maintaining control over information than a declaration of the vice president in June 2007. At that point, Vice President Cheney declared his office to be exempt from the executive branch information reporting requirements on the grounds that the Office of the Vice President is not part of the executive branch. This argument was based on the idea that, as the titular president of the Senate, the vice president has responsibilities both in the executive and legislative branches, so is part of neither. While this argument is virtually without merit in terms of the Constitution—the vice presidency is unambiguously listed as part of the executive branch in the Constitution and has been a part of the executive branch for the history of the republic—it is indicative of a broad picture of misuses and abuses of the structure, history, and integrity of the republic itself in service of the Bush administration's goal of controlling information. It also reaffirmed the lengths to which the Bush administration went to achieve its aims of information control.

   Taken as a whole, the Bush administration matched or outperformed the worst excesses in U.S. history by the federal government in terms of limiting both access to information and individual rights. In fact, many of the information policies of the Bush administration echo the bleakest episodes of executive misuses of information policy in U.S. history. Historical differences in levels of suppression of dissent depend heavily on "the extent to which national political leaders intentionally inflamed public fear" (Stone, 2004, p. 533). Even the greatest presidents in U.S. history have taken significant steps to curtail rights in reaction to major external threats. However, the Bush administration acted as though these dark episodes in the use of presidential power were benchmarks to exceed. Consider the following historical parallels.

   In a time of fear of war with one or more of Britain, Spain, and France, the Congress passed and President John Adams signed into law the Alien and Sedition Acts, intended to significantly curtail the ability of citizens to speak against the policies and activities of the government. The Alien and Sedition Acts comprised the Naturalization Act, the Alien Friends Act, the Alien Enemies Act, and the Sedition Act; together, these acts set limits not only on speech, but also on definitions of personhood and citizenship (Lewis, 2008). The acts were quickly rejected by the citizenry and the courts—and soon after, the legislature—as being antithetical to the Constitution, and particularly the First Amendment. Similar laws were created during World War I. The Alien and Sedition Acts of 1917 and 1918 made pacifism and opposition to the war illegal, and allowed the post master the authority to ban seditious mails. The laws were also used to prosecute and harass labor unions, radicals, homosexuals, labor activists, any protestors, and anyone who criticized the president (Brinkley, 2008).

The Bush era brought a series of laws, like the USA PATRIOT Act and the Homeland Security Act, as well as activities like warrantless wiretapping programs and massive database mining, that cut significantly into the scope of the First Amendment by allowing the government to monitor what citizens read, say, and do. The creation of these laws and programs was sufficient to chill much free expression, but the capabilities of modern ICTs meant that the government could collect amazing amounts of information about the beliefs and expressions of virtually every citizen (Bamford, 2008). Bush Justice Department memos coming to light after the administration left office included assertions that the First and Fourth Amendments could be unilaterally suspended by the president, who was also deemed able to order the use of the military against targets within the United States (Smith & Eggen, 2009). In terms of the theory of information worlds, such claims are particularly disquieting examples of the powers of social typing, since they explicitly define a particular social type—the president—as the sole arbiter not only of rights related to information but also of other fundamental rights.

During the Civil War, a battle for the very existence of the republic, President Abraham Lincoln suspended the right of habeas corpus for prisoners of war. This action was decried by many critics at the time and stands as the single smudge on the reputation of Lincoln who safely guided the republic through its darkest days. It also occurred in a context where abolitionists had significant constraints placed on their speech, as they faced suppression of their mailings, gag rules on debates, riots in the North and South, and seditious libel prosecutions (Lewis, 2008). During the course of the war, Lincoln pursued prosecutions of newspaper editors, politicians, and individual citizens who objected to the war, even issuing a warrant for the arrest of the chief justice of the Supreme Court that was never executed, while in the confederacy, very similar clampdowns on speech were occurring (Lewis, 2008). The Bush administration worked for the passage of the Military Commissions Act, which not only suspended the right of habeas corpus much more broadly but also codified many other limitations on the rights of prisoners—foreign or domestic—to know the information being used against them or to have access to counsel or many other central aspects of the judicial process in the United States.

In one of the most graphic examples of restrictive uses of social typing in the history of the country, after the United States was pulled into World War II, President Franklin Roosevelt created internment camps for Japanese-Americans. These citizens were herded into the camps for the duration of the war, many losing their property, and had no right to appeal their imprisonment. The Bush administration created Gitmo in Guantanamo Bay, Cuba—a prison treated as being completely free from the burdens of American or international law—as well as the dark site prisons of the Central Intelligent Agency (CIA) in Eastern Europe, which were far more secretive than Gitmo. They then created the extraordinary renditions program to kidnap suspects and remove them to the custody of other nations with far fewer legal protections than the United States provides.

The Bush administration also upped the ante considerably by authorizing the defiance of the Geneva Conventions—the use of techniques such as waterboarding that have been among favored torture practices of despotic governments dating back to the Spanish Inquisition—in the handling of these prisoners to get them to reveal information (Shane & Mazzetti, 2009). Less than two months after the Bush administration left office, a federal probe revealed that the CIA had videotaped ninety-two harsh interrogations of prisoners, including acts of waterboarding and other methods of torture; the tapes were destroyed by the CIA before the investigators for the federal probe could see them (Johnson & Warrick, 2009). These same techniques had been deemed illegal in the United States since the 1890s, and the use of such techniques had resulted in executions for war crimes after World War II.

The value of torture as a means of gaining information was so engrained in the beliefs of the information world of the Bush administration that former Secretary of State Condoleezza Rice, in trying to defend the use of waterboarding, echoed the language that Richard Nixon used to defend the Watergate break-in and cover-up—when the president, as a uniquely positioned social type, defined as the unitary executive, authorizes an activity, it is "by definition" not illegal. The former vice president went several steps further in early 2009, going on a virtual speaking tour of talk shows to enthusiastically proclaim the information value of waterboarding and other forms of torture; in the first six months after the 2008 presidential election, he made more talk show appearances than he did in any year he was in office (Balz, 2009).

During the height of the Cold War in the 1950s and 1960s, the director of the FBI and the attorney general oversaw a program that had the FBI spying on approximately 10 million Americans, usually because they believed in causes—such as desegregation, feminism, pacifism, and antifascism—that the executive branch disagreed with or found politically inconvenient. At the height of these activities, the FBI not only monitored the information behavior of average citizens involved in political causes, but also journalists, members of Congress, congressional staffers, antiwar protesters, student groups, university faculty, and the Democratic Party, often using many illegal information collection techniques like warrantless wiretapping and searches (Stone, 2004, 2008). As an example, the FBI tracked Albert Einstein for decades without a warrant—reading his mail, repeatedly searching his home and office, recording his activities and travels, and cataloging his guests, among many other intrusions—because he was a prominent activist for civil rights and international peace (Jerome, 2002). These activities were supported by numerous citizens' organizations, many of which were founded (often funded or encouraged by the government) to promote patriotism during these years. Groups such as the American Protective League, the National Security League, the Knights of Liberty, the American Defense Society, and the Boy Spies of America spied, eavesdropped, intercepted mails, reported people to the government, conducted raids, and made citizens arrest; the largest was the American Protective League with over 250,000 members (Brinkley, 2008). The actions authorized by the Bush administration in the name of the "war on terror" used all of the techniques employed by the programs of the 1950s, but with extremely more effective ICTs and a radical expansion of both the kinds of information being collected and the numbers of people whose personal information was collected.

Although the actions of the Bush administration, thus, mirror similar practices of preceding administrations, modern ICTs have provided the means to go much further in the denials of constitutional rights. And—as probably goes without saying—information is at the heart of each of these issues today. Further, previous efforts by presidents to severely limit information access have led to individual government agencies actually exceeding the limitations to ensure compliance; thus, policies limiting access to information set by a president serve as a baseline for such limitations, not the uppermost limit (Hogenboom, 2008). The members of the small world of the Bush administration were even apparently aware of their historical antecedents. John Yoo (2008), one of the architects of the Bush administration's anti-terror policies, has written as a defense of their extra-constitutional excesses that at least they were not as bad as FDR creating detention camps.

The Bush administration added its own layers of innovation by creating new ways to limit information access—scrubbing websites, reclassifying previously publicly available information, creating dozens of new undefined designations to keep information secret, pulling countless records from libraries and archives, redacting government publications, greatly expanding the number of people who had the authorization to classify information, stacking scientific panels and the entire Department of Justice with people who shared their norms and goals, changing the content of government reports, and creating new protections to keep presidential and vice presidential records from ever becoming public, among many other activities that significantly affect information access and exchange (Jaeger, 2007).

And these mostly were accomplished through executive orders, presidential signing statements, agency memos, and other methods that are de facto law-making that does not involve the legislative process but, once again, situates the president as the sole arbiter of information value and information behavior (Gorham-Oscilowski & Jaeger, 2008; Halstead, 2008; Tatelman, 2008; Wang, 2008).

It is temptingly easy to look at the various attempts to control information as a series of unique cases and events, missing the larger picture. Entire books have been written about secrecy and lack of transparency in the Bush administration (Gup, 2007; Roberts, 2006). However, the issues go beyond individual means of controlling information. The sheer volume of major changes to information policy, types of information held secret, and modifications to the constitutionally defined protections for information in society is stunning, collectively far surpassing anything undertaken by previous administrations in terms of placing radical constraints on the ability of people to access and use information.

And yet, even in light of the consistent actions of the administration between 2001 and 2009 and the resulting negative impacts on information available for public discourse and government decision-making, the assertion by the Office of the Vice President that it was not part of the executive branch was startling. The idea that the fundamental structure of the republic—the existence of three separate branches of government—would be undone to avoid an information-reporting requirement shows the extraordinary length to which the administration was prepared to go in order to promote the views of its small world. Cheney's argument implied that the vice presidency enjoys the privileges and protections of the executive branch, yet exists entirely outside of the reporting requirements of either the executive or the legislative branch. In 2001, in relation to his Energy Task Force meetings, Cheney already rejected any membership in the legislative branch, as his argument for withholding that information was that it was privileged executive branch information beyond the reach of the legislature (Relyea & Halchin, 2003). Cheney's attempt to establish his own personal fourth branch came, coincidentally, at the same time that the *Washington Post* published an exposé on his activities that revealed that he had a safe in his office large enough for a person to fit in to keep his everyday paperwork and that he had invented a designation for literally all of his documents to be stamped—"Treated as Secret/SCI" (Becker & Gellman, 2007a, 2007b; Gellman & Becker, 2007a, 2007b).

Cheney's argument that the vice presidency existed outside of the three branches of government came only after he tried unsuccessfully to close the office responsible for collecting the information he wanted to hide. The fourth branch assertion, however, is a bit different from all that came before. Previous Bush administration efforts to control information had been to remove it from public view, prevent it from ever getting before the public, or to badly distort information that the public managed to get access to. All of the efforts to control information revolved around the information itself and the structures for its dissemination, access, and exchange. This fourth branch maneuver, however, casually threatened to undermine the most basic of foundational elements of the American republic for the sake of maintaining control of information (Jaeger, 2009b). It is extremely revelatory regarding the lengths that the administration was willing to go to preserve and expand its control over information. The small world of the Bush administration learned to use information as a very effective weapon. It is hard to imagine a greater sign of disregard for the Constitution and the idea of democracy itself than to try to alter the established structure of government to avoid releasing information to another government agency.

To claim a part of the federal government is not part of any of the branches—if even in terms of information requirements—ultimately could open a Pandora's Box, as other government functions or agencies that could be deemed to fall conveniently outside of the three branches. Simply for the sake of furthering their control over information, the Bush administration imagined and tried to implement an internal threat to the function of the government in a way that no previous administration had dared. And now that this idea has been suggested, who knows how it may be revived by future administrations wishing to exert similar domination over information.

SMALL WORLD INFORMATION POLITICS

Information policy under the Bush administration was strongly modified in favor of an information value devoted to controlling information, keeping it away from the public, other branches of government, and even government agencies within the executive branch. However, particularly in an era defined by the speed and capacity of technology to support widespread and open access to and exchange of information, "information access and exchange lies at the heart of deliberative democracy" (Jaeger & Burnett, 2005, p. 474); in such a context, these policies raise enormous questions about the health of the republic, especially if the policies are continued by subsequent administrations. When information is not available for scrutiny, analysis, and dialogue by the public, the press, and other branches of government, and when the government collects more information about citizens while releasing less information about its own activities, the actions of the executive branch can occur unchecked by the delicate democratic balance envisioned by the framers of the Constitution.

As a potential counterbalance to the information politics like those of the Bush administration, certain public sphere entities have staunchly defended the principles of information access. Traditional entities like libraries have worked aggressively to resist the information controls and information collection programs that the administration has implemented. The very public crusades of public libraries to resist the limitations on free access to information have simultaneously caused libraries to be viewed negatively by some in government—best demonstrated by then-Attorney General Ashcroft's frequent criticism of librarians in response to their resistance to the USA PATRIOT Act—and very positively by much of the public. Standing up against encroachments on information rights has been a longstanding library belief, and one that served to increase public trust in libraries (Jaeger & Fleischmann, 2007; McClure & Jaeger, 2008b). The small worlds of librarians have served to balance administration policies and raise public awareness of these activities throughout information worlds.

Another kind of resistance to government control of information can be found among the previously described social networks found on the Internet. Newer entities of the expanded landscape of technologically supported information worlds have worked to provide free access and exchange through online information channels like e-mail, websites, wikis, and blogs. These entities have significant amounts of social capital and are strongly trusted by their members and communities. Although their political power and access to "resources that can be used for influencing the choices of voters and, between elections, of officials" (Dahl, 1961, p. 4) is limited when compared to the federal government, as their membership is decentralized and generally less able to act on the political stage, there are still some examples of small worlds outside the confines of government structures and mass media outlets that have used online information channels to achieve great success in sharing the information value of their world far beyond the boundaries of the world. Perhaps the most notorious of such a small world outlet having a dramatic impact both on the broader lifeworld and the workings of the government itself is the 1998 report of then-President Bill Clinton's affair with White House intern Monica Lewinsky, which remained untouched by more established media until it was broken by the Drudge Report, a blog devoted to aggregating news stories and, in particular, emphasizing conservative interests (BBC News, 1998).

Overall, the decentralized nature of the Internet and the accompanying social networks can have a very significant impact not just on information exchange, but on the entire political processes that shapes information worlds. The organizing capacities of the Internet are often viewed as an effective means to overcome the disparities rooted in unequal distribution of access to information and the power to disseminate it. However, the political value of the Internet depends on the way it is used and by whom, with much of what is perceived as the creation of new social capital and new forms of involvement actually lacking any meaningful deliberative component (Thompson, 2006). Generally, the ways in which it has been used by political parties and other powerful information worlds means the Internet has not fostered deliberation, but encouraged political parties in many countries to take the easier and more populist route of catering their stances to press-friendly issues and popular opinions, heavily based on the rapid electronic feedback via news sites, party sites, and online political forums (Rogers, 2004).

Early studies of the use of the Internet by political parties show how rapidly the online environment has become central to formal and informal political organization. In 1999, party presence could be measured in terms of such basic measures as having "web sites listed in Yahoo" and references to sites in print newspapers and magazines, online news sites, and Usenet groups (Margolis, Resnik, & Wolfe, 1999, p. 28). In addition, the early history of presidential candidates in the United States and the Internet has hardly been smooth. In 1996, Bob Dole became the first U.S. presidential candidate to mention his website in a debate, but he said the wrong address and the people who eventually found his correct URL went to a crashed website. In a 2000 debate, Al Gore famously made the awkward assertion that he "took the initiative in creating the Internet," a statement that makes much more sense in light of Gore's active promotion of the policy shifts that resulted in the opening of the Internet to commercialization in the mid 1990s—which did, indeed, help to create the Internet as it is today—than it does from any technological point of view. Prior to the 2008 election, successes with online fundraising in presidential races were unusual, being an aspect of otherwise poorly funded and unsuccessful campaigns—McCain and Bradley in 2000; Dean in 2004 (Klotz, 2004).

In contrast, the Obama campaign in 2008 launched its own social networking site—MyBarackObama.com—the day the campaign was announced. Over the course of the twenty-one month campaign, two million profiles were created on the network, 400,000 blog posts were written, 200,000 offline events were planned by users and 35,000 volunteer groups were created (Vargas, 2008a). The campaign also created profiles on sixteen social networking sites, posted 1792 videos to YouTube, uploaded over 50,000 photos to Flickr and "tweeted" over 200 times on Twitter (Owyang, 2008). More than 13 million e-mail addresses were collected by the campaign, and 7,000 different targeted messages were delivered for a total of more than one billion e-mails sent (Vargas, 2008b). More than one million voters signed up to receive text message updates, receiving anywhere from five to twenty targeted text messages per month based on state, region, zip code, and college campus affiliation (Vargas, 2008b). Largely as a result of these Internet-based efforts, the Obama campaign raised over $600 million—more than $500 million of that came from donations made online, with 3 million donors making a total of 6.5 million donations online (Vargas, 2008b). Since taking office, President Obama has appointed a chief information officer and a chief technology officer, and strongly encouraged government agencies to use Facebook, YouTube, Flickr, Ning, Twitter, and other social networking applications to disseminate government information and directly communicate with citizens (Choney, 2009).

The Internet also gives greater voice to parties or groups that are less well known and less mainstream in their values. Perhaps not surprisingly, minor political parties in industrialized nations embraced the Internet as a means of reaching voters much faster than major parties did (Klotz, 2004). Some of these groups hold very extreme political positions; already by 2002, there were 25,000 known hate group and terrorist websites (Klotz, 2004). "The success of the interest group propagandist depends in no small measure on his skill in selecting the appropriate media and channels for his efforts at communication" (Truman, 1971, p. 244). For both mainstream and fringe groups, the Internet has been incredibly helpful at linking people with like-minded persons into small worlds. Among both conservative and liberal blogs, for example, more than 90% of links are to like-minded sites (Sunstein, 2008).

These uses of the Internet to harness the power of small worlds to affect discourse and policy in information worlds can have very positive impacts, such as the large level of involvement in the political process by otherwise less involved small worlds by the Obama campaign. However, it can also, through processes of social typing and polarization, lead members of deliberating groups to solidify their own biases and values, resulting in more extreme positions. This outcome can occur for issues of fact as well as opinion (Brown, 1985; Turner, 1987). In many cases, if people are deliberating with people of similar opinions, "their views will not be reinforced but instead shifted to more extreme points" (Sunstein, 2002, p. 186). Typically, group polarization not only increases extremism, it makes individual members of the group significantly more homogenous, squelching diversity (Sunstein, 2008). And the result of polarization is a reduction in communication and interaction, a key example of which can be found in the polarization of elected officials in the U.S. federal government. "Both parties have used hardball tactics that have polarized Washington, with Democrats and Republicans alike punishing members who cross party lines and rewarding loyalists with generous campaign contributions" (Eilperin, 2006, p. 5).

Low-choice media environments encourage centrism and reduce polarization because there are so few outlets through which to be seen (Prior, 2007). However, the explosion of potential information sources online and through other technologies can exacerbate the opportunities for small worlds to engage in group polarization, altering patterns of involvement in larger information worlds. "Greater media choice has made partisans more likely to vote and moderates more likely to abstain. Politics by choice is inherently more polarized than politics by default" (Prior, 2007, p. 263). However, for all of these concerns, the social networks of the Internet have already shown the ability to compete with the narratives of mainstream political groups.

In the 2007 campaign for prime minister of Australia, the Australian media—much of which is owned by Rupert Murdoch—openly supported the conservative party and its prime minister, going so far as to selectively report and distort the results of their own polls, particularly those of the major papers owned by Murdoch (Bruns, 2008). As a result, blogs and other online social networks played a large part by providing contrasting views to balance the media coverage. "Months of persistent efforts by bloggers and citizen journalists in Australia to neutralize and counteract news media industry spin in political reporting left leaders of the journalism industry in an uneasy jittery mood" (Bruns, 2008, p. 66). In this situation, the small world controlling the mainstream media was pitted against the many small worlds present online, with control of discourse about an election in the information world of a nation at stake. Based on the outcome of that election, the small worlds online apparently made a stronger case.

Entire political movements now exist and sustain themselves through the capacities of the Internet to disseminate information. In the aftermath of the Iranian election of 2009, for instance, during the final days of the preparation of this book, some of the potentials of an ICT like Twitter became clear. Even though a relatively small number of people in Iran used it to post small bits of information about protests in the streets of Tehran (whether through text, photographs, or short videos), Twitter still became one of the primary means through which the world outside of Iran learned about those protests. In this case, the apparent technical limitations of the tool—only 140 characters per post and the consequent focus of individual "Tweets" on minutia—and the fact that Twitter relies on decentralized distribution of messages combined to make it an ideal way for protesters to side-step Iranian efforts at censorship and make information about events in their world available around the globe (Cohen, 2009).

Perhaps the most durable of these Internet-based political movements is one that has garnered a significant amount of attention—the Zapatista Army for National Liberation (EZLN). The EZLN was formed in the early 1980s by a small number of neo-Marxist intellectuals and activists who moved into the mountains of Chiapas to organize for the rights of indigenous populations who had been alternately neglected by or forced to assimilate by the Mexican government (Barmeyer, 2009; Gilman-Opalsky, 2008). When the movement began to encourage uprisings in the mid-1990s, the Mexican government worked to portray them as simple terrorists, while the EZLN tried to rely on disseminating information to convince the international community that it was an oppressed rights organization that required safety and solidarity (Gilman-Opalsky, 2008). The EZLN was more successful, receiving support from more than 140 non-governmental organizations (Hanson, 2008).

The EZLN began to distribute texts that held sway with the indigenous people they were fighting for, the general public they needed sympathy from, and the international governments they were seeking support from. "Although the political texts of the rebels are principally for the masses, they seem well thought out and written to seduce or irritate an elite" (Paz, 2004, p. 33). By engaging both the small worlds of the indigenous people and the lifeworld of greater Mexican culture, they were able to use information to suffuse their beliefs throughout the information worlds of their society. They also timed release of their texts to maximize awareness of other small worlds and the lifeworld. For example, their first major declaration was issued the same day the North American Free Trade Agreement (NAFTA) came into effect (Hanson, 2008).

As a result, the EZLN created a transnational public sphere that was "a place for the formation of will and opinion" and "committed to the broadcasting of already-formed opinion and will to other publics and to power holders" (Gilman-Opalsky, 2008, p. 248). This public sphere rapidly moved from printed materials and television exposure when it could be had to the Internet and the omnipresent ability to disseminate political information. Their listserv has been operating without interruption since 1994 (Jung, 2003). The movement also crafted and marketed an appealing image to a range of small worlds around the globe—by turns threatening, poetic, humorous, and mysterious, with a clear visual image—that thrived through electronic media, counterbalancing the social type of simple terrorists that the Mexican government wished to portray them as (Bob, 2005; Gilman-Opalsky, 2008). Amazingly, for all its distasteful militaristic and revolutionary aspects and activities, the Zapatista movement has increased focus on the rights of indigenous people not only in Mexico but throughout Central and South America, and inspired a range of international solidarity groups. One international supporter of the ELZN cause described their communications as having "read like poems and conveyed an urgent immediacy that was hard to resist" (Barmeyer, 2009, p. x).

The binding thread of these groups is a focus on indigenous rights from a region-specific perspective (Gilman-Opalsky, 2008). All of this has been cultivated through "words, images, imagination, and organization" spread via the Internet (Cleaver, 1998, p. 81). Websites and listservs have been established by supporting organizations around the world, including ones at several universities in the United States, to provide the EZLN texts and messages in many different languages (Hanson, 2008). By framing the problems of one small cultural group in one nation as a symbol of the negative impacts of globalization on many small worlds in different places, the EZLN became an inspiration for and symbol of certain parts of a global justice movement (Bob, 2005). Rather than being utterly ignored or eradicated by the Mexican government as a minor irritant, the Zapatista movement successfully spread its message across an increasing number of small worlds, ultimately influencing the discourse at the national level in Mexico and at the international level.

In sharp contrast to the Bush administration's approach of placing radical constraints on the availability of information throughout the lifeworld, the EZLN has used information for political gains by building upward one small world at a time. In both cases, a particular information world has tried to extend its own internal understanding of information value to the broader lifeworld. Both have been successful, but in very different ways. The top-down movement of the Bush administration was based on the political force of the government over the lifeworld of the nation and, in some cases, other parts of the world. In sharp contract, the EZLN has survived by building from the bottom up, spreading its information value across the boundaries of many different, but similarly disenfranchised, small worlds that collectively have made an impact on the discourse in the lifeworld.

INFORMATION WORLDS IN THE POLITICAL WORLD

The information value that ultimately takes prominence will significantly shape the ways in which information functions in democratic societies, which in turn will determine how democratic societies really can be. The ancient idea of the polis encompassed three concepts—the political community comprised by the citizenry, the public space necessary for citizens to communicate, and the issues of public concern (Brenkman, 2007). While contemporary ICTs open up new ways for the public, public spaces, and public concerns to come together, these same technologies offer new ways for governments to limit information access.

If the restrictions on information behavior in either the lifeworld or in small worlds are too great, the public sphere will not be able to function effectively, and information worlds will not be able to contribute meaningful discourse. "A healthy public sphere is both necessary for, and an indication of, the exercise of the rights of free speech and of participatory citizenship" (Green, 2001, p. 116). If a society's public policy, the actions of the government, and the actions of corporations constrict the public sphere too tightly, thereby limiting the information behavior in either small worlds or the lifeworld, the democratic nature of the society is jeopardized.

However, the small worlds in a democratic society will often not oppose limitations of their rights, particularly if they are afraid. The efforts of the Bush administration proved rather successful at parlaying fear into acceptance of limitations of rights. For example, a great deal of effort was put into creating a sense of fear about cyber-attacks. The idea of massive, crippling cyber-attacks on society have penetrated the popular imagination, the political discourse, and the reports of the media, but "despite the persuasiveness of threat scenarios, cyber-threats have clearly not materialized as a 'real' national security threat" (Cavelty, 2008, p. 4). It also has been suggested that people are compensated for the loss of their own rights and liberties at home in exchange for the spectacle of military might and success leading to the loss of rights and liberties of people in other countries (Pease, 2003). While there are large differences between negative freedom (the freedom from . . .) and positive freedom (freedom to . . .), social turmoil can make it difficult for citizens to understand the relationships between the two (Berlin, 1969).

However, another aspect of the acceptance of limitations of rights may be tied to the complexity of governments and an accompanying lack of government information at this point. "Americans must now cope with a political system that works in opaque and mysterious ways that probably no one adequately understands" (Dahl, 1994, p. 13). The additional layers of government created by the homeland security state may serve to alienate many citizens from their sense of their own individual rights. Relics of the homeland security era—such as the five-color threat level index that, based on its usage, seems designed to always remain at least at the yellow, elevated threat level—often proved hard to locate within reality. When the information world running the government controls the threat level that a country faces, it means that a single world literally is in charge of the threat and how it is perceived. "The proclamation of the threat level is an admission that there is no threat. Or if a threat exists, the government is powerless to deal with it" (Mamet, 2004, p. 6). Terminological changes like these across the lifeworld of a society will certainly affect the smaller information worlds it contains.

The term "homeland security" was never used before 9/11 in the United States, and even contradicts the terminology previously used to describe the country, which reflects the serious shift in the government's understanding of information value that it accompanied. "The phrase 'Homeland Security' . . . is confected and rings false, for America has many nicknames. . . . But none of us *ever* referred to our country as The Homeland" (Mamet, 2004, p. 6). The term "homeland security" was, in fact, first used by the apartheid-era South African government (Maxwell, 2005). Homeland evokes the imagery of slavery or forced immigration—the country left behind often involuntarily and that one might return to one day—not the country where one still resides (Mamet, 2004; Pease, 2003). "The homeland engendered an imaginary scenario wherein the national people were encouraged to consider themselves dislocated from their country of origin by foreign aggressors so that they might experience their return from exile in the displaced form of the spectacular unsettling of homelands elsewhere" (Pease, 2003, p. 8).

Yet, many other nations have followed the lead of the United States and embraced the concept of homeland security, leading to significant changes to the information policy environment, as well as the creation of new policies related to information collection, access, and security (Caidi & Ross, 2005; Hosein, 2004; Ross & Caidi, 2005). As a result, the perspectives of the information world of the Bush administration have stretched across information worlds, both small and large, around the globe.

When policy impacts a lifeworld, it also impacts the constituent small worlds and how they use information, ultimately affecting the public sphere. Public sphere entities can protect and reinforce information behavior in a lifeworld and small worlds, but these protections can be limited by the colonization of the lifeworld through government or corporate intrusions. Such intrusions have become more sophisticated as ICTs have become more pervasive in everyday life. "[T]he body politic becomes more fragile with every separation of decision and participation, every undoing of the checks and balances, and every divergence of word and deed" (Brenkman, 2007, p. 14).

Policy decisions about communication and information—information access, freedom of expression, intellectual property, privacy, regulation of media, physical communication structures and the support of education, research, and innovation—all shape the parameters and the health of the public sphere in a society (Starr, 2004). The amount of information that is available to develop, articulate, and communicate opinions on social and political issues is key to democratic participation and to the functioning of the public sphere. "Public affairs in a democracy is, among other things, a stream of collective consciousness in which certain actions . . . come to be noticed and remembered. . . . They are observed by politically aware citizens trying to size up events in their environment" (Mayhew, 2000, p. 5). To put it in the terms of the theory of information worlds, the most important information value in a democracy is the broadly agreed-upon belief that open access to information is not only a virtue but a necessity if a democratic information world is to succeed, as are information behavior that emphasize attention and action.

In considering the relationships between public policy and the public sphere, the United States provides an informative example. In the United States, an active public sphere has existed as long as the republic as a result of freedom of the press and an open society: "attentive strata of the public and opinion fluidity has been there from the start" (Mayhew, 2000, p. 8). Under policies promoting freedom of access to information and the provision of forums for discussion in the public sphere, democratic government flourished in the United States. Though the general trend was toward continually increasing information access and exchange through public policy for most of the republic's history, a significant shift may have occurred in the policies that shape the public sphere (Jaeger & Burnett, 2005). While it is true that "To deny the term *democracy* to any regime not fully democratic in the ideal sense would be equivalent to saying that no democratic regime has ever existed" (Dahl, 1982, p. 7), too much control of information by one very powerful small world can be catastrophic for the health of the public sphere as well as that of all of the information worlds that feed it.

Or perhaps these concerted attempts to control information behavior will ultimately be undone by Bush's replacement. Within the first week of the Obama administration, President Obama rescinded President Bush's Executive Order 13233 and issued the Presidential Memorandum on Transparency and Open Government, measures designed specifically to rein in the information control excesses of the Bush years. In addition, President Obama established the http://www.change.gov site to get ideas and feedback from citizens and mobilize their community participation. In combination, these actions can have a substantial impact on the availability, access, and dissemination of government information, as well as the ability of the public to interact with government (Bertot, Jaeger, Shuler, Simmons, & Grimes, 2009). The Obama administration, however, also apparently will follow some of the leads of the preceding administration—a June 2009 news conference by President Obama included prearranged questions from two invited reporters who were not normally at White House press conferences (Milbank, 2009).

The Bush-era limitations have affected information behavior in many public sphere entities, including public schools, public libraries, universities, the news media, and online. The online environment, with its uniquely diffuse nature—its ability to link members of small worlds across great distances, to expose members of small worlds to the perspectives of many other small worlds, and to allow specific small worlds a forum to articulate their own opinions—offers perhaps the greatest hope for a public sphere entity that can continue to cultivate access to and exchange of political and social information across information worlds, regardless of policy intentions.

9 Applications of the Theory of Information Worlds

As the previous chapters have laid out the concepts, ideas, and potential areas of application for the theory of information worlds, this chapter will wrap together the ideas introduced in the book and examine the future directions of these concepts, including ways in which the theory of information worlds can be used in different research contexts. A key part of this chapter is the analysis of the research questions and opportunities raised by the concept of information worlds for LIS and other disciplines that study the social and political aspects of information. This chapter also reflects on the ways in which the theory of information worlds can serve as a theoretical driver across disciplines to advance understandings of the social and political roles of information in society.

Ultimately, a theory will only be of genuine scholarly and social value if it is used and modified through application in various scholarly contexts. By examining the connections and potential avenues of research for information worlds in various contexts, the chapter will hopefully serve as a starting point for the further development of the theory of information worlds. To further this research, each of the key contexts discussed in previous chapters—the conceptual (information value and access), the social (libraries and other public sphere entities), the technological (the Internet and Web 2.0), and the political (the media and governance)—will be examined as sources for future scholarly work.

INFORMATION WORLDS REVISITED

Prior to delving into the various research possibilities presented by information worlds, a short summary of the key concepts of the theory of information worlds seems in order. While the preceding eight chapters have provided an examination of the concepts in significant detail, recapping the key concepts will help to approach the discussion of potential future applications of the theory.

In sum, the theory of information worlds is designed to provide a framework through which the multiple interactions between information,

information behavior, and the many different social contexts within which it exists—from the micro (small worlds) to the meso (intermediate) to the macro (the lifeworld)—can be understood and studied. The theory argues that information behavior is shaped simultaneously by both immediate influences, such as friends, family, co-workers, and trusted information sources of the small worlds in which the individual lives, as well as larger social influences, including public sphere institutions, media, technology, and politics. These levels, though separate, do not function in isolation, and to ignore any level in examining information behavior results in an incomplete picture of the social contexts of the information. As such, scholarly explorations of information behavior must account for the different levels to fully understand the social drivers of information behavior and the uses of information in society.

Building primarily upon the theoretical works of Habermas and Chatman, the theory of information worlds seeks to expand the perspective brought to studies of information behavior in society and to increase understanding of the myriad ways in which information plays a significant role in social, political, and personal lives. While Habermas focused on the large social structures of the lifeworld and Chatman focused on the individualized units of small worlds of society, the theory of information worlds argues for the examination of information behavior in terms of the small worlds of everyday life, the mediating social institutions of the public sphere, the lifeworld of an entire society, and the hegemonic forces that shape the lifeworld. These social structures constantly interact with and reshape one another. In examining these interrelated parts, the theory of information worlds focuses on five social elements:

- Social norms: a world's shared sense of the appropriateness of social appearances and observable behaviors
- Social types: the roles that define actors and how they are perceived within a world
- Information value: a world's shared sense of a scale of the importance of information
- Information behavior: the full range of behaviors and activities related to information available to members of a world
- Boundaries: the places at which information worlds come into contact with each other and across which communication and information exchange can—but may or may not—take place

As with the social structures within information worlds, the elements are interrelated and constantly interact with and influence one another.

As localized information worlds, each small world has its own social norms, social types, information behavior, and understanding of information value. The members of each small world have established ways in which information is accessed, understood, and exchanged within their world and the degree to which it is shared with others outside the small world. Few individuals, however, exist only in one small world; it may not even be possible except in extreme circumstances of social isolation. In contrast, there is no real limit to the number of small worlds to which an individual can belong. A typical person is a part of many small worlds—friends, family, co-workers, people with shared hobbies, etc.

Any one of these small worlds may offer many places where its members are able to interact with members of other small worlds. Information moves through the boundaries between worlds via people who are members of two worlds or through interaction between members of two small worlds in a place where members of different small worlds are exposed to other perspectives. Further, the contact between small worlds and other inputs from the lifeworld can lead to the creation of new worlds as information passing over the boundaries between worlds either blurs those boundaries or otherwise transforms or changes information behaviors and perceptions of information value. Encountering other small worlds can occur through public sphere institutions, such as in a public library, or through new technological avenues of communication and exchange, such as social networks on the Internet. As information moves through boundaries between small worlds, the information is treated, understood, and used differently in each small world in line with the social norms of that world. As a result, the same information may have a different role within each small world.

Together, these small worlds constitute the lifeworld of information. The way that the small worlds in the lifeworld as a group treat information will shape how the information is treated within the lifeworld as a whole. As the information moves between small worlds, more and more small worlds will decide how to treat this information, generating an overall perception of the information across the lifeworld. The more small worlds that are exposed to information the more exchange between small worlds there will be, and the better chances there will be for a democratic perception of and approach to the information.

However, beyond the small worlds, there are also influences at play in the lifeworld which shape the way that small worlds perceive information. Some of these influences increase contact between small worlds and promote democratic engagement in the lifeworld.

Libraries, schools, and other public sphere organizations exist specifically to ensure that information continues to move between the small worlds and that members of each small world are exposed to other small worlds. In sharp contrast, other influences serve to constrain the movement of information between small worlds or constrict the socially acceptable perceptions of information. The most influential information worlds—such as those who possess political power or those who control the media—can use their power to push back against the collective small worlds to enforce a minority perception on the majority, asserting control over the information in the lifeworld.

Some influences on small worlds and the lifeworld are inherently neutral but can be used for the objectives of either increasing or decreasing information access and exchange. ICTs act as a way for small worlds to connect in new ways and to reach other small worlds that would not otherwise touch their boundaries. The Internet and online social networks represent particularly powerful examples of this phenomenon. But ICTs—like the Internet and more traditional media—can also work to homogenize perspectives or enforce hegemonic perspectives of small but powerful small worlds on the lifeworld. In total, the small worlds are shaped by all of these larger influences, but also have the power collectively to define the parameters of the external influences.

The theory of information worlds, thus, attempts to account for all of these social and structural elements at work in the shaping of information behavior within a society. While there is obviously great benefit in studying the ways in which information behavior is shaped by the micro, the meso, or the macro level, studying them across levels will provide a much richer and nuanced understanding of the ways in which information is perceived and moves though society. Though the theory of information worlds presents a much more complex approach, it is intended to compensate the researcher by providing a more thorough and realistic picture of the issues being studied.

CONCEPTUAL APPLICATIONS OF THE THEORY OF INFORMATION WORLDS

In developing the theory of information worlds, the intent was to create a theory that did not deal with information as an isolated concept. Too often, information is relegated to being an unspoken and unexamined part of social theories; alternately, it is treated as the only aspect of a theoretical problem. Along with attempting to craft a means by which to explore the multiple, interrelated levels of information in society, the theory of information worlds is meant to bring together LIS and elements of social theory, public policy, communication, media studies, computer science, education, computer-mediated communication, and other areas of research that are essential to understanding information as a social and societal issue. Similarly, the incorporation of these elements of other disciplines is intended both to make the theory more effective and to make it useful in many different disciplines.

A single event in any information world represents many opportunities for research for these different disciplines. Consider an event where the information is of importance to the lifeworld, to those information worlds in power, and to every other small world in a society—a large-scale economic crisis that has a strong negative impact on the global economy, perhaps. In such a situation, information related to the crisis would be created, disseminated, accessed, and exchanged very rapidly across small worlds and throughout the lifeworld. The social norms, social types, information behavior, information value, and boundaries of each small world would shape a reaction to the information. In the case of a stock market crash, small worlds comprised of individuals of higher socio-economic status would likely give much more thought to the information than individuals from small worlds of lower socio-economic status. In contrast, massive job losses in blue collar sectors would likely be information of much greater interest to small worlds of lower socio-economic status than small worlds of higher socio-economic status.

Depending on which small worlds are most concerned, ICTs may play different roles in the crisis. If the crisis was of the stock market crash variety, the blogs, wikis, and online social networks would likely be ablaze with an indistinguishable rush of rumor, falsehood, fact, and supposition about the crisis. Similarly, the cable news channels—particularly the channels that cover financial news—would be denominated by up-to-moment market numbers and wild speculation about what to do with money based on the social norms and information value of the small world of network market analysts. In short, there would be no shortage of information for concerned small worlds to process, though much of it would be harmful and contradictory. The overabundance of information might also lead investors from interested small worlds to act when they might otherwise not, encouraged by the chattering classes of the media and the rumor and sigh of the Internet.

If the crisis were of the massive blue collar job losses type, phone and face-to-face communication—and e-mail to a lesser extent—would likely serve as a primary means of information sharing among those affected. The interactive online tools would play much less a role in information sharing, as those of lower socio-economic status are less likely to use the Internet, have Internet access, or to use advanced online tools (Horrigan, 2008). The cable news channels would also be a less vibrant source of information about job losses, as they reflect the interests of members of small worlds that watch their programming and the interests of the small worlds of the owners and employees of the channels. It has been hard to miss the fact that the real economic crisis of 2009 has caused channels like CNBC and Fox Business to argue strongly for bailouts for corporations and tax cuts for the wealthy, while arguing as strongly against bailouts for individual mortgage holders and tax increases for the wealthy.

The information worlds with political power will also have important reactions to this financial information, depending on their political persuasion. The Republican Party, for example, has spent the past three decades building its economic policy entirely around the idea of continual tax cuts for the highest income earners as the best means to spur the economy (Bartels, 2008; Hacker & Pierson, 2005a, 2005b, 2007; Hamburger & Wallsten, 2006; Jacobs & Shaprio, 2000). As result, information about any economic downturn will be filtered through this lens. Further, with the traditional ties of the Republic Party to corporations, if the information worlds controlling political power were Republican, the stock market crash would likely receive more attention. If the worlds with political power were Democratic, in contrast, the information about the job losses would likely receive more attention, given the traditional interests and affiliations of that party.

In either of these events, public sphere agencies would likely try to provide information to help those affected. In the current economic downturn in the real world, use of the public library for job-seeking activities, social services, e-mail access, entertainment, and other purposes has skyrocketed (Carlton, 2009; Van Sant, 2009). Further, as many people consider home Internet access to be a luxury that can be cut to save money in harsh economic times (Horrigan, 2008), this usage of libraries for information access and exchange is likely to continue to increase.

Such an economic scenario would provide numerous options to study the roles of information through the theory of information worlds from varying scholarly perspectives—LIS, public policy, computer-mediated communication, political science, and journalism, among others. The theory of information worlds not only offers an opportunity to examine information at the conceptual level but also provides means to study the large- and small-scale pictures of how to improve information flows in practical circumstances, by illuminating potential differences between the norms, values, and behaviors of those in different worlds, allowing information services and policy makers to make more informed decisions.

PRACTICAL APPLICATIONS OF THE THEORY OF INFORMATION WORLDS

A clear example of the ways in which the theory of information worlds can be used as a means to conceptualize and analyze information problems and offer practical solutions is in the context of emergency situations. During an emergency, accessing, exchanging, and implementing information is of enormous importance for responders and for residents affected by the emergency. Key roles of information in emergencies include:

- Alerting first responders
- Coordinating emergency response
- Notifying community members
- Gathering facts about the situation
- Connecting family and friends
- Identifying those in need of assistance
- Sharing updates about conditions (Jaeger, Fleischmann, Preece, Shneiderman, Wu, & Qu, 2007)

However, moving information in emergencies is extremely difficult for a number of reasons.

There are tens of thousands of governmental and non-profit emergency response organizations in the United States (Pelfrey, 2005; Wise, 2006). "The disaster network is loosely structured, organizationally diverse, motivated by a broad range of interests, and in part ad hoc" (Waugh & Sylves, 2002, p. 148). For example, 1,607 governmental and non-governmental organizations were involved in the response to 9/11 terror attacks in New York City (Kapucu, 2004). During 9/11, responders from different New York City organizations were unable to communicate due to incompatible radio systems, as the Fire Department was still using the same analog radios that failed in the same way during the response to the 1993 attack on the World Trade Center (Kettl, 2004). Further, ICTs used by residents often fail during emergencies, as was demonstrated by failures of landlines, cell phones, and the Internet due to the infrastructure damage and volume of people attempting to use them (Dearstyne, 2007; Will, 2001).

During the response to Hurricane Katrina, the response problems were due mostly to lack of information flows. Federal, state, and local government agencies and private organizations did not know what actions to take in the response, did not have any guidance on how to coordinate and interrelate their activities, lacked an overall operational concept, and had no system to track and share information (Comfort, 2007; Garnett & Kouzmin, 2007; Wise, 2006). Secretary of Homeland Security Michael Chertoff told Congress the response was "significantly hampered by a lack of information on the ground," while the White House report on the failures of the Katrina response observed that an "inability to connect multiple communication plans and architectures clearly impeded coordination and communication at the federal, state, and local levels" (Chertoff, 2005; White House, 2006). Later investigations, however, have strongly suggested that government officials had access to large amounts of information about the extent of the situation and in many cases failed in inter-agency coordination and communication of information (Bier, 2006; Brinkley, 2006; Cooper & Block, 2006; Dyson, 2005). The problems with information led to delay, duplication, lack of coordination, and confusion that ultimately proved extremely deadly (Dearstyne, 2006).

Clearly there are organizational and technical challenges to the availability and movement of information in emergency response that the theory of information worlds cannot help to address. However, the preparation for and response to emergencies can be been seen as a cycle with information sharing being key throughout the cycle (McEntire, 2002; Pelfrey, 2005). Studies have repeatedly demonstrated the difficulties of coordination between responders, residents, government agencies, businesses, volunteers, and relief organizations in an emergency (Haddow & Bullock, 2003; Jones & Mitnick, 2006; Kapucu, 2004; McEntire, 1997, 2002; Portsea, 1992). "Sharing information, willingness to collaborate, and shared values" are vital bases of effective information sharing and communication in major disasters (Kapucu, 2004, p. 210). In crisis situations, insufficient or incorrect information often leads to either complete inaction or disastrous action (Dearstyne, 2005, 2006). As a result, information access and exchange among residents and responders are among the most pressing issues in an emergency, an area where research using the theory of information worlds may be of great assistance.

A primary approach in terms of the theory of information worlds would be studying ways to improve the exchange of information between the information worlds of different government and non-profit responders, between small worlds of different residents, and between small worlds of responders and residents. Examining the respective social norms, social types, information behavior, information value, and boundaries of two small worlds of responders (say, the local police and fire departments of a community) could help to identify better ways to exchange information between those two small worlds. By examining enough of these different small worlds, a better understanding of information exchange and use between small worlds and across the lifeworld could be developed. If members of different worlds understand each other's norms, behaviors, and perceptions of information value, they will be better situated to collaborate and to share information across the boundaries of their worlds.

The theory of information worlds could also be used to help understand the best ways to increase trust of emergency information. A wide range of factors, including language barriers, cultural differences, trust of government, and socio-economic status, can make certain groups harder to reach with emergency information (Dyson, 2005; Lindell & Perry, 1992; Tierney, 2006). However, successful risk communication is based most importantly on contextualizing and personalizing the risk to engender trust (Lindell & Perry, 1992). In high-stress situations, people base the majority of their credibility assessments on how trustworthy a source is—not the level of expertise (Covello, 1996). Individuals who trust a source find its risk estimates more credible and its policies more acceptable (Johnson & Slovic, 1995). By focusing on the social norms, social types, information behavior, information value, the boundaries of small worlds that do and do not trust emergency information, the specific sources that are trusted, and the reasons for that trust, the means to engender trust of emergency information and sources among less trusting small worlds could be devised. Answering these types of trust questions would help to reduce scenes of people who refuse to prepare for or evacuate before pending emergencies.

Some of these methods by which to increase trust of emergency information in small worlds and across the lifeworld could be learned from studying the success of one specific public sphere agency in becoming a trusted source of emergency information. Through the devastating 2004 and 2005 hurricane seasons, public libraries became widely trusted sources of emergency information to prepare for, respond to, and recover from hurricanes

(Bertot, Jaeger, Langa, & McClure, 2006a, 2006b; Jaeger & Fleischmann, 2007; Jaeger, Langa, McClure, & Bertot, 2006). "In those emergency situations, public libraries were able to serve their communities in a capacity far beyond the traditional image of the role of libraries, but these emergency response roles are as significant as anything else libraries could do for their communities" (McClure & Jaeger, 2008b, p. 88).

Indeed, one of the authors of this book had a student in his online class whose home was destroyed by Katrina in the middle of the semester. Because of her local library, where she worked, not only was she able to access the technology that allowed her to complete the semester, but she was also able to play a practical role in the recovery of her own local community and in similar towns across the Gulf Coast. This student, one could say, successfully melded the multiple small worlds of which she was a part—the worlds of student, library employee, and disaster victim—leveraging them to achieve things that no single world by itself could have made possible.

The combination of localized services, assistance, and support provided by public libraries made them perhaps the most trusted source of emergency information in the public sphere during those hurricane seasons. Studies of the ways in which such trust was cultivated could also be significant in establishing a better understanding of the ways information is used and exchanged in emergencies. Studies of the worlds of librarians could also be very revealing in terms of the lessons they learned through performing these roles, the ways in which the members of this world feel information could better reach other worlds, and the norms and information behavior that developed among librarians in response to these new social roles and expectations.

Another aspect of information worlds that would be significant in such investigations would be the movement of information through the traditional media and through the Internet. Several traditional media channels exist for disseminating emergency information, most notably the radio- and television-based Emergency Alert System. Local television stations also provide coverage when possible. These methods of information dissemination, however, may not be trusted in all small worlds, and they only provide one-to-many communication—there is no multi-directional information exchange. Media that are used to share information in an emergency will only be effective to the extent that people have access to the technologies being used to share information (Burkhart, 1991). It is extremely important to make the emergency information that traditional media provide trusted across all small worlds. Otherwise, as more interactive technologies like the Internet provide faster and broader emergency information, small worlds with the money to afford computers and Internet access will be privileged over other small worlds in their awareness of and ability to respond to an emergency.

For those with access to the Internet at home or at work, Internet-based communities have proven very effective at prompting access, exchange, and dissemination of emergency information between small worlds. During recent emergencies such as the Asian tsunami and the California wildfires of 2007, blogs, wikis, and other tools were used to spread up-to-moment information, offer assistance, deploy volunteer resources, and coordinate community responses (Jones & Mitnick, 2006; Palen, Hiltz, & Liu, 2007). Further, some news organizations have harnessed the interactive power of the Internet to create new channels for emergency exchange of information, leveraging the capacities and prevalence of mobile telecommunications devices to involve residents in the process of information gathering and reporting, not only integrating readers' voices into the news via in-house news blogs and wikis but also providing more information and footage than a news organization could gather on its own (Masie, 2005). The day after Hurricane Katrina made landfall, at any given moment "there were hundreds of Citizen Journalists feeding content in real time" through cell phones, mobile devices, and computers (Masie, 2005, p. 74). These individuals were able to bring information from their small worlds and present it directly to the lifeworld. Research based on the theory of information worlds could explore how to maximize the use of the Internet and Internet-based social networking tools to promote the flow of information between small worlds and into the lifeworld.

In total, efforts such as these could lead to much better understandings of how to access, exchange, and disseminate information in emergency contexts to help both those affected and those responding. It could also increase the trust of such information, as well the efficiency and effectiveness with which it is moved through information worlds. By focusing on this vital emergency problem at multiple levels and by analyzing the information flows between information worlds, through public sphere agencies, through the traditional media, through Internet-based social networks, and across the lifeworld, the concepts of the theory of information worlds could help to provide very practical solutions to the problem of emergency information. And emergency information is but one of many potential areas where the theory of information worlds could be applied by researchers to address significant problems.

RESEARCHING INFORMATION WORLDS

Building on the preceding discussions of conceptual and practical applications of the theory of information worlds, this section approaches conceptual and practical cross-disciplinary research areas raised by the theory of information worlds. These potential research areas are examined in terms of each of the lenses through which the theory was presented in preceding chapters. While the preceding sections demonstrated the potential uses of the theory beyond the issues detailed within the arguments of this book, the goal of this section is to offer suggestions for further explorations of the ideas detailed in the book. This discussion is not intended to provide a list of extremely specific topics but instead to explore general areas that contain many potential research topics.

Though the theory of information worlds relies heavily on the concept of information value as an element of the theory, information value also offers research applications for the theory. One area in which the theory of information worlds could clearly be applied is in the exploration of the ways in which different small worlds balance the elements of information value— perception, content, control, access, benefit, utility, format, and commodity. The ways in which these elements are balanced heavily influences the access, exchange, and use of information in personal, social, corporate, and governmental contexts.

Research applications of the theory of information worlds in the context of information value present opportunities to study single small worlds, interactions between multiple information worlds, or the entire lifeworld. Perceptions about the appropriate use, value, and appropriate forms of access to information will shape views about levels of openness, privacy, secrecy, and control that are acceptable in society and in individual lives. As a result, perceptions of information value shape the courses of information that will be trusted, the means of access that will be trusted, and the ways in which information will be shared. All of these issues present significant research opportunities.

Further, information value is not necessarily tied to authority or accuracy or consistency or subject domain of information. Tensions often arise between the information value ascribed to known sources and perhaps more authoritative sources, as well as between competing known or authoritative sources. Such tensions between different information values of different small worlds can heavily define information worlds. These tensions can also serve as the basis of research applications of the theory of information worlds.

Information access also presents a range of research opportunities. Each type of information access—physical, intellectual, and social—provides research avenues by which the theory of information worlds could be applied. As technological change greatly expands the number of ways to access information and reduces the usage of some longer standing means of access, the approaches of different small worlds to physical information access, their social norms and information behavior related to their chosen means of access, and the impacts on technologically disadvantaged small worlds all present areas for applying the theory of information worlds in relation to physical access.

Intellectual access presents issues within the small worlds themselves, such as the study of the ways in which a small world establishes frames through which an individual views information that has been accessed. Social norms, social types, and information value will have no small impact on intellectual access; in turn, these impacts on intellectual access shape how the individual will understand the public sphere and other aspects of the lifeworld around them.

Research problems focusing on social access will primarily relate to the ways that information moves within a small world and across the boundaries between small worlds. The social norms, social types, information behavior, and information value of a small world will frame how a piece of information is viewed and exchanged within the small worlds, as well as the way it is or is not shared with other small worlds. Social access issues will greatly influence the approach of a small world to certain information. Collectively, the social access of many small worlds will impact the role of that information within the public sphere. Applications of the theory of information worlds to the study of all types of access could provide many insights into the roles of small worlds within information worlds.

The agencies of the public sphere—particularly public libraries—will continue to play socially significant roles regardless of technological change. As more information has become available through the Internet, people have turned to public libraries to help them get access to the Internet, learn how to use it, and receive guidance on finding the most relevant and authoritative information online. Beyond these new technologically enhanced roles, public libraries continue to serve as a place where small worlds gather, interact with one anther, and are exposed to information and perspectives from many different small worlds and from across the lifeworld. As demonstrated by the spike in library use during the global recession of 2008–2009, the traditional and technological roles of libraries have merged seamlessly in their work as the most venerable pillar of the public sphere.

These key roles of the library as a public sphere agency—providing information and a means for small worlds to interact—raise numerous potential applications of the theory of information worlds. One of the most important applications may be in the study of the reasons for the public libraries' continued success, while other traditional aspects of the public sphere have disappeared or grown severely weakened in social influence. Similarly, use of the theory to help understand the ways in which libraries can continue to provide these vital contributions to the lifeworld and to individual small worlds will be of great value. At the highest level, the study of public libraries could even be framed in terms of the study of the health of the public sphere as an entity.

Research applications of the theory of information worlds could have a great deal of practical consequence. Public libraries, in spite of their enormous social contributions, are perpetually underfunded. This situation is due in no small part to insufficiencies of the basic counting metrics of resources usage to telling the whole story of the contributions that libraries make. Perhaps research based on the theory of information worlds could provide a richer, more detailed portrait of the contributions of libraries beyond simple metrics that could ultimately lead to greater support for public libraries and their sustenance of the public sphere.

Of the topics discussed in this book, the Internet and online social networks present perhaps the widest canvas in which the theory of information worlds can be applied. Given the potentially ever-expanding nature of the online world, coupled with the movement of more and more educational, professional, governmental, commercial, and recreational activities into cyberspace, the relationships between the multiple and ever-increasing worlds of the Internet offer practically limitless opportunities to study information behavior and the social and contextual value of information.

The geography-defying connections of the Internet allow for the creation of previously impossible small worlds and allow existing small worlds to become exposed to many, many other small worlds that they would never otherwise encounter. As a result, each small world online has the ability to identify and explore boundaries between a limitless supply of other small worlds. It also allows for the ability to broadcast information almost instantaneously across small worlds. These highly unique aspects of the Internet mean that information will be accessed, exchanged, and used at enormous speed through a wide array of small worlds—both those that have offline counterparts and those that do not—reframing and reinterpreting the information to establish the parameters of how the information will be used in the lifeworld of the Internet.

While many users of the Internet stick exclusively to their own small worlds and the information that is acceptable within these worlds, the increasing use of social networking technologies continues to make it harder to avoid exposure to information from outside the boundaries of particular small worlds. As a result, the theory of information worlds seems able to offer many avenues by which to examine online information behavior.

One of these potential avenues involves an issue mentioned in passing several times: the role of popular culture. Numerous online worlds have sprung up over the years to support a variety of fan communities, with interests ranging from books for youth such as the Harry Potter series to rock bands like the Grateful Dead. The theory of information worlds suggests that at least some of the information of most interest—or the most value—within these worlds may be, because it focuses on pop culture phenomena and entertainment, considered lacking in importance by members of other worlds. The theory of information worlds can provide a framework through which to investigate not only the significance of such seemingly trivial information to members of specific worlds, but also the ways in which it is intertwined and enmeshed with political, social, or other information coming into the worlds from beyond their boundaries.

Topics related to traditional broadcast news and the media, on the other hand, present what may be some of the most far-reaching research applications of the theory of information worlds. The theory presents a framework by which to try to analyze the movement of information throughout the lifeworld and into individual small worlds as driven by traditional media. While the movement of information through social networks and other less formal online information channels are faster, the traditional media still maintain a significant amount of gravity and the capacity to broadcast very widely. In spite of the inroads of the Internet into the normal flow of information, an extremely large number of small worlds continue to prioritize information provided by the broadcast network and cable news networks.

Given their dramatic reach and ability to interject information into the lifeworld and countless small worlds simultaneously, the applications of the theory of information worlds to traditional news and broadcast media may best begin with examinations of the hegemonic influence of these organizations over many parts of the information worlds. The interrelationships between the media and political life in the public sphere offer many major research possibilities, as do the efforts of the media to move into the online environment to extend their influence, even through the acquisition of social networking sites that were initially designed to go around the influence of the mainstream media.

Shifting perceptions of the traditional news and broadcast media among small worlds, the ways in which small worlds balance information from the media and from other sources like online social networks, and information behavior of small worlds that ignore the media entirely all present compelling avenues to apply the theory of information worlds in the media context from the perspective of individuals and small worlds. Exploring the media from the perspective of information worlds also provides opportunities to examine the ways in which changing perceptions of the media by different small worlds indicate the future levels of influence that the media may have over the lifeworld.

The theory of information worlds opens new possibilities for understanding public policy and political processes. With perceptions of information so vital to the success of policy and political initiatives, an understanding of political information flows though and between small worlds and across the lifeworld would provide new insights into the political dimensions of societies. It could also help to illuminate the ways in which different small worlds become involved in or disengaged from political information. Given the high levels of community engagement by some small worlds and a sheer lack of interest in engagement by others, this problem is especially valuable for research to consider using the frame of the theory of information worlds.

As the discussions in Chapter 8 suggested, the influence of political forces on information worlds at all levels can be generated from the utmost hegemonic elements of society, as in the case of the information world of the Bush administration, or can be generated by the disenfranchised and oppressed elements of society, as evidenced by the EZLN. These two radically different social positions reveal the breadth of ways in which politically motivated information worlds can influence discourse in the lifeworld and spread their message among small worlds. The new capacities for technology in this process, as evidenced by the overwhelming success of the online fundraising and coordinating efforts of the Obama presidential campaign, provide further opportunities for the application of the theory of information worlds in the context of the political world.

Finally, only limited mention of the meso level of information worlds between the micro and the macro has been made in this text, as the micro and macro levels provide the clearest examples through which to explain the theory. The theory of information worlds, however, could provide a valuable framework for understanding how information works and is exchanged in such settings, providing a lens through which to analyze how corporate and organizational structures impact information use. Further, it could enhance the understanding of how individuals and small groupings of individuals function and interact when they are members not only of their own small worlds but also of the meso worlds in which they work and the macro worlds that provide the broadest context for their lives.

Ultimately, the most useful and important applications of the theory of information worlds may fall entirely outside the areas discussed in this chapter. The discussions herein are by no means intended to be limiting. Any ways in which these ideas can be applied in research in any field will contribute to discourse about the roles of information in society. The greatest concern for social theory of information is not establishing limitations, but instead developing a richer, more dynamic body of theory. The final chapter of the book explores this problem and the place of theory of information worlds within this context.

10 The Future of Information Theory

The 1966 publication of the first volume of *Annual Review of Information Science and Technology* in many ways began the process of formalizing the study of information as a scholarly field (Cuadra, 1966). During the late 1960s, professional and scholarly literature began to try to articulate a vision of the then new concept of "information science" (Borko, 1968; Miles, 1967; Taylor, 1966). These authors generally framed information science in terms of "properties and behaviors of information, the use and transmission of information, and the processing of information for optimal storage and retrieval" (Borko, 1968, p. 4). The social issues related to information typically were not emphasized in these discussions, while theory was acknowledged as having a role in informing practice, but in the most general terms. Unfortunately, the vagueness about the role of theory—including social theory—in the study of information issues has been perpetuated in the ensuing four decades.

Critiques of the insufficiency of theory in LIS are far from new (e.g., Budd, 1995; Buschman, 2006; Harris, 1986). Clearly, there are LIS scholars working on theory development, including social theory about information. However, the key issue is that there are far too few scholars working on issues of information theory and not enough connections between theory, research, and practice in LIS.

Part of the difficulty of building a greater role of social theory in the study of information derives from the historical antecedents of the field of LIS, particularly the ways in which the field of library education was originally conceived and developed. LIS education as a formalized endeavor was well established before the close of the nineteenth century, but the educational environment was comparatively limited, deriving from early library education leader Melvil Dewey's beliefs that LIS education was strictly practical training (Wiegand, 1996). In the early years of LIS education, many LIS schools were actually affiliated with libraries and staffed by practitioners rather than scholars (Houser & Schrader, 1978). The first doctoral degree in LIS was not created until 1928 when the University of Chicago's program finally brought LIS education to the level of a research discipline, though LIS education of all types was still strongly professional.

The significant influence of the 1923 Williamson report initiated the process of making LIS education more intellectual by bringing scholarship and scholars into the previously intensely practical degree programs (Houser & Schrader, 1978). By the mid-twentieth century, faculty in LIS schools typically possessed both a doctorate and master's degree in LIS (Weech & Pluzhenskaia, 2005).

The move from a traditional library focus to a broader information focus significantly expanded the scope of the pedagogy and the backgrounds of the faculty members in LIS (Dillon & Norris, 2005; Weech & Pluzhenskaia, 2005). Information science has been described as "a field of professional practice *and* scientific inquiry" (Saracevic, 1999, p. 1055). Yet, a late-1990s analysis of the educational culture of LIS suggested that the field suffered from being highly resistant to change and recommended embracing interdisciplinary education and the development of new specializations among programs (Sutton, 1998; Van House & Sutton, 1996). A 1995 study of LIS doctoral programs found evidence of increasing emphasis on both quantitative and qualitative research methods, interdisciplinarity, technology-based research, and interaction with related fields (Powell, 1995). However, LIS has traditionally had a very strong predisposition to predictive means of inquiry and education (McGrath, 2002).

By the 1990s, LIS programs widely acknowledged the importance of social institutions in the delivery of information, though there remained major differences in the perceived pedagogical and research goals of LIS (Miksa, 1991; Pawley, 2005). An early 1990s study asserted an educational core for LIS comprised of information organization, information systems, users and access, evaluation, and information management (Drabenstott & Atkins, 1996). A study later in the 1990s suggested that LIS education was not unified by core concepts, but instead defined by a user-centered perspective and a focus on the cognitive and social aspects of information (Pettigrew & Durrance, 2000). Markey's 2004 study of the trends in LIS education suggested there are fairly standard core elements across programs. However, information theory has not been a part of these attempts to define the parameters of LIS.

Even the recent banding together of the leading research-oriented LIS programs in North America as the "iSchool movement" has not brought greater emphasis to theory in the field. The iSchools define themselves as "schools interested in the relationship between information, technology, and people" (http://www.ischools.org), and the iSchools include programs that focus on information science, library science, and communication, usually with programs in at least two of these areas (Mokros, 2008). However, in spite of the expanding scope and interdisciplinarity, they have yet to identify information theory as a core part of their mission and endeavors.

The increasing emphasis on a broader information focus in LIS programs has resulted in strong criticisms of LIS education over the past few decades (Dillon & Norris, 2005). In 2004, former ALA president Michael

Gorman pointed out that more "library-oriented" programs were "under assault from 'information science,'" and these programs were devoted to "abstract theory, especially in areas only marginally related to librarianship" (Gorman, 2004, pp. 99–100). Similarly, many information professionals criticize the scholarly literature for being too laden with theory (Powell, Baker, & Mika, 2002). It is, indeed, interesting to note the degree to which such a simmering controversy in LIS mirrors some of the tenets of the theory of information worlds. The appropriate role of theory in LIS education, because it is largely a matter of differing perceptions of information value across the small worlds of LIS educators (and there may be several such worlds), is contested. Some educators—members of one small world—place a high value on theory-driven approaches, while others criticize it for being too removed from the nuts and bolts of the practice of librarianship. Further, as anybody who has attempted to incorporate theory of any kind into teaching LIS courses can attest, students all too often perceive theory as a pointless abstraction, taking them away from what they consider to be truly useful procedural and hands-on training. While such criticisms are based on little fact, the introduction of theoretical components to teaching and research in LIS clearly is still strongly resisted by some factions of the field.

For example, the perceived overabundance of theory in LIS claimed by Gorman, sadly, is mostly the product of his imagination. A 2002 study found that only 34.2% of articles in leading LIS journals discuss theory, and those papers discussing theory were often written by scholars educated in fields outside of LIS (McKechnie & Pettigrew, 2002). Not surprisingly, the focus on increasing usage of theory in LIS often turns directly toward applying theories from other fields. For example, when Wiegand (1997) issued a challenge to expand the use of theory in library research, papers answering his call tended to assert that the work of some particular theorist—most specifically Bourdieu, Foucault, Gramsci, Habermas, and Marx—could be applied to library research (e.g., Budd, 2003; Buschman, 2006; Raber, 2003; Radford, 2003). Though not responding to Wiegand's call, one of this book's authors has also suggested that a specific outside theorist's work—in this case, Paul Ricoeur's hermeneutics—could be appropriately imported into LIS research (Burnett, 2002).

A further challenge is that many scholars in LIS still equate "research with evidence gathering at the methodological level," placing greater emphasis on data collection than on the larger goals of research, such as what they want to know and why those data are needed (Hernon & Schwartz, 2009, p. 31). However, such unease about developing a native theory may not be atypical among interdisciplinary fields. The search for legitimacy by interdisciplinary fields—described as "anxiety discourse"—is tied to fears of irrelevance and intellectual shortcomings (King & Lyytinen, 2004). However, an interdisciplinary field can strive to be a setting for intellectual and methodological diversity that accounts for and fosters otherwise unlikely exchanges and advancements from interactions between the range of areas of study within the field (Al-Hawamdeh, 2005; Lyytinen & King, 2004).

This brief overview of the growth pangs of theory in LIS points to the fact that the very field devoted to the study of information has yet to come to terms with, much less give concerted effort to focus on, the development of information theory through the course of nearly a century and a half. Coming to terms with the intellectual consequences of theory can induce fear of uncertainty (Bauman, 1992). Perhaps LIS, as a field, has a pronounced fear of the unknown.

It has been suggested that "the creation of meaning through interpretation of social action and relations can be seen as one of the main prerogatives of contemporary scientific research" (Hansson, 2005, p. 111). There is no reason that other fields cannot also work toward the development of a richer body of social theory of information, but it seems that the field most naturally positioned to take the leadership position in such endeavors, in fact, has not yet prioritized theory development. This situation is particularly ironic given that libraries are highly valued for serving as the marketplace of ideas, but scholars seem reticent to really explore how the marketplace functions at a conceptual level.

In terms of the theory discussed in this book, the small world of LIS scholars has not given great value to social theory of information, resulting in norms and behaviors that rarely focus on the theoretical aspects of their work. Through the boundaries with related scholarly small worlds, this lack of interest in theory has shaped how other fields approach information, ultimately limiting the social theory of information within the greater information world.

As has been noted, LIS approaches to information theory remain rare, at best, and research still is strongly tied not only to the practical origins of the field, but also to the scientific understanding of theory, in which the value of theory is tied exclusively to its predictive power and is measured by the strictly controlled testing of hypotheses. Even an LIS scholar as closely associated with qualitative research methods as Elfreda Chatman followed such a course, presenting her theories with sets of formal predictive propositions. By contrast, the path that has been pursued here is more closely akin to the place of theory in philosophical and humanities-based disciplines. Though it provides a robust framework by which to analyze and understand information in society, the theory of information worlds is less narrowly predictive than it is a multifaceted conceptual framework. It is the authors' hope that, by presenting a set of concepts and analyses related to the social role of information in a wide variety of contexts and worlds, this book can play a small role in furthering the dialogue about the relationship between information and society.

In 1831, the great abolitionist activist and author William Lloyd Garrison wrote, "I have need to be all on fire, for I have mountains of ice about me to melt" (quoted in Meyer, 1998, p. ix). While arguing for the expanded use of theory in the study of information hardly rises to the same level as the fight for equality for all people, there is something of the same need to be a staunch and incessant advocate to try to change attitudes that are deeply engrained and at times surprisingly antagonistic. Yet despite the resistance to social theory of information in many scholarly quarters, the need for information theory designed to examine the social roles of information becomes more pronounced with each technological leap that makes information more an integral component of the structures of society.

In the age of the Internet, information has become the lifeblood of government, commerce, education, social interaction, and many other daily activities. In most societies, information behavior now defines the rhythms of everyday life. "We are constantly in dialogue and our position becomes transformed by the process of understanding" (Cornelius, 1996, p. 19). Truly, there is a desperate need for information theory to help to identify, understand, contextualize, and enrich the place of information across all levels of society to improve the lives of individuals, the discourse online and in the public sphere, the vitality of democratic participation, and many other social processes. For scholarship to best meet the needs and reflect the realities of society, the discourse about social theories of information must be vibrant, robust, and dynamic.

A text on the philosophy of librarianship that was originally published in French in 1976, but only translated into English in 2009, demonstrates how long the need for greater information theory has been recognized and how slowly action has been taken: "Unlike most other social sciences, which are primarily concerned with theory, librarianship has mostly developed its practical side, to the neglect of its theoretical side" (Cossette, 1976/2009, p. 37). Both the positives and negatives of the progress that has been made in the development of information theory can be seen in a recent book of brief essays on theories and concepts of information behavior, which discusses more than seventy approaches that can be used to study information behavior, though the preponderance of these are concepts that focus on the micro level and about half are taken from other fields (Fisher, Erdelez, & McKechnie, 2005). Another edited volume from 2009 focuses on the application of theoretical constructs to the intersections of libraries and technology, though the majority of constructs discussed in the text are direct applications of ideas of theorists outside LIS—Karl Marx, Max Weber, and Habermas—or the application of the conceptual frameworks developed by other fields—technological utopianism, feminism, and technology as ideological phenomenon (Leckie & Buschman, 2009). While both of these edited volumes engage theory in LIS areas of research, both also suggest a clear hesitation to develop a robust, large-scale engagement with theory in the native terms and frameworks of LIS.

While many fields need to be a part of the development of more social theory about information, LIS—the field that is inherently devoted to information—should put great effort into taking the lead in the advancement of these theories. And the first step is for LIS to cultivate more native theory. These theories can be built upon a foundation of elements from LIS or from other fields or from both, with the theory of information worlds being an example of the third option. LIS scholars must also put greater effort into addressing the hesitations to more extensively engage with theory as a field. This hesitation limits the development of theory, impedes the creation of connections between theory, research, and practice, and marginalizes LIS discourse from the discourse of other fields. It also ultimately serves to prevent LIS from reaching its potential as a body of scholarly work.

The theory of information worlds, regardless of the flaws and weaknesses that will need to be corrected as the theory matures, can serve as a part of the solution to this problem. But no single theory will be able to address the vast complexities of information in its myriad roles across the layers of society. Along with presenting one theoretical frame for better understanding information behavior in society, the theory of information worlds will also hopefully help to encourage the discussion, use, and development of theory in this vital area. A greater effort toward theory building is clearly needed in LIS, not only as a field ideally positioned to develop theoretical approaches to information in society but as a means to greatly assist its practitioners through the development of more theoretical supports for practice in the information professions. Along with LIS, all fields that deal with information in society—communication, computer science, public policy, and human computer interaction, among others—must commit themselves to a stronger focus on the theoretical dimensions of information behavior in society. Information is established as a major driver of all levels of social interaction, and scholars must work to ensure that theoretical research meets the needs and realities of information-enabled individuals and societies.

Bibliography

Ackermann, K. D. (2007). *Hoover, the red scare, and the assault on civil liberties.* New York: Carroll & Graf.

Aitken, R. (2007). *Can we trust the BBC?* London: Continuum.

Alejandro, R. (1993). *Hermeneutics, citizenship, and the public sphere.* New York: State University of New York Press.

Alexander, L., Carter, D., Chapman, S., Hollar, S., & Weatherbee, J. (2008). MLibrary 2.0: Create, share, and network. *College & Research Libraries News, 69*(4), 204–206, 229.

Alfino, M., & Pierce, L. (1997). *Information ethics for librarians.* Jefferson, NC: McFarland & Company.

Al-Hawamdeh, S. (2005). Designing an interdisciplinary graduate program in knowledge management. *Journal of the American Society for Information Science and Technology, 56*(11), 1200–1206.

Alliance Library System. (2008). *Trends report.* Available: http://www.alliancelibrarysystem.com/pdf08/TrendsReport2008.pdf

Allner, I. (2004). Copyright and the delivery of library services to distance learners. *Internet Reference Services Quarterly, 9*(3), 179–192.

American Library Association. (1939/1948). *Library bill of rights.* Chicago: Author.

———. (1995). *Code of ethics.* Chicago: Author.

American Library Association & Information Institute. (2007). *Libraries connect communities: Public library funding and technology access study 2006–2007.* Chicago: American Library Association.

American Library Association Washington Office. (1988). *Less access to less information by and about the U.S. government: A 1981–1987 chronology.* Washington, DC: Author.

———. (1992). *Less access to less information by and about the U.S. government: A 1988–1991 chronology.* Washington, DC: Author.

Anderson, B. (2007). Social capital, quality of life, and ICTs. In B. Anderson, M. Brynin, J. Gershung, & Y. Raban (Eds.), *Information and communication technologies in society: E-living in a digital Europe* (pp. 162–174). London: Routledge.

Anderson, B., Brynin, M., Gershung, J., & Raban, Y. (2007). *Information and communication technologies in society: E-living in a digital Europe.* London: Routledge.

Anderson, N. (2006). Tim Berners-Lee on Web 2.0: "Nobody even knows what it means." *Ars Technica News.* Available: http://arstechnica.com/business/news/2006/09/7650.ars

Appiah, A. (2005). *The ethics of identity.* Princeton, NJ: Princeton University Press.

Asch, S. (1952). *Social psychology*. Englewood Cliffs, NJ: Prentice Hall.

Aschmann, A. (2002). Providing intellectual access to cooperative extension materials. *Quarterly Bulletin of the International Association of Agricultural Information Specialists, 47*(3/4), 89–92.

Augst, T. (2001). Introduction: Libraries and agencies of culture. In T. Augst & W. Wiegand (Eds.), *Libraries as agencies of culture* (pp. 5–22). Madison: University of Wisconsin Press.

Bain, R. (1943). Sociometry and social measurement. *Sociometry, 6*(3), 206–213.

Baker, N. (1996). The author vs. the library. *New Yorker, 72*(31), 51–62.

———. (2001). *Double fold: Libraries and the assault on paper*. New York: Random House.

Baldasty, G. J. (1993). The rise of news as a commodity: Business imperatives and the press in the nineteenth century. In W. S. Solomon & R. W. McChesney (Eds.), *Ruthless criticism: New perspectives in U. S. communication history* (pp. 66–97). Minneapolis: University of Minnesota Press.

Balz, D. (2009, May 14). As Cheney seizes spotlight, many republicans wince. *Washington Post*, A1, A4.

Bamford, J. (2008). *The shadow factory: The ultra-secret NSA from 9/11 to the eavesdropping on America*. New York: Doubleday.

Barber, B. R. (1994). *Strong democracy*. Berkeley: University of California Press.

Bardoel, J., & Lowe, G. F. (2007). From public service broadcasting to public service media: The core challenge. In G. F. Lowe & J. Bardoel (Eds.), *From public service broadcasting to public service media* (pp. 9–28). Goteborg, Sweden: Nordic Information Centre for Media and Communication Research.

Barker, A. N. (2005). Executive Order No. 13,223: A threat to government accountability. *Government Information Quarterly, 22,* 4–19.

Barlow, J. P. (1996). A declaration of the independence of cyberspace. Available: http://homes.eff.org/~barlow/Declaration-Final.html

Barmeyer, N. (2009). *Developing Zapatista autonomy: Conflict and NGO involvement in rebel Chiapas*. Albuquerque: University of New Mexico Press.

Barnum, G. (2002). Availability, access, authenticity, and persistence: Creating the environment for permanent public access to electronic government information. *Government Information Quarterly, 19,* 37–43.

Bartels, L. M. (2008). *Unequal democracy: The political economy of the new gilded age*. New York: Russell Sage Foundation.

Bauman, Z. (1992). *Intimations of postmodernity*. London: Routledge.

Baym, N. K. (1995). The emergence of community in computer-mediated communication. In S. G. Jones (Ed.), *CyberSociety: Computer-mediated communication and community* (pp. 138–163). Thousand Oaks, CA: Sage Publications.

BBC News. (1998, January 25). Scandalous scoop breaks online. Available: http://news.bbc.co.uk/2/hi/special_report/1998/clinton_scandal/50031.stm

Beck, U. (2002). The terrorist threat: World risk society revisited. *Theory, Culture and Society, 19*(4), 39–55.

Becker, J., & Gellman, B. (2007a, June 26). A strong push from backstage. *Washington Post*. Available: http://blog.washingtonpost.com/cheney/chapters/a_strong_push_from_back_stage/index.html

———. (2007b, June 27). Leaving no tracks. *Washington Post*. Available: http://blog.washingtonpost.com/cheney/chapters/leaving_no_tracks/index.html

Becvar, K. M., & Srinivasan, R. (2009). Indigenous knowledge and culturally-responsive methods in information research. *Library Quarterly 79*(4), 421–441.

Bednarek, M. (1993). Intellectual access to pictorial information. *Australian Library Journal, 42,* 33–46.

Bell, L. (2008). *HealthInfo Island final report*. Available: http://www.alliancelibrarysystem.com/pdf08/healthinfoislandfinalreport08.pdf

Bennett, S. (2001). The golden age of libraries. *Journal of Academic Librarianship, 27*, 256–259.

Berlin, I. (1969). *Four essays on liberty.* New York: Oxford University Press.

———. (1996). *The sense of reality: Studies in ideas and their history.* New York: Farrar, Strauss & Giroux.

Berman, H. J. (1983). *Law and revolution: The formation of the Western legal tradition.* Cambridge, MA: Harvard University Press.

———. (2003). *Law and revolution II: The impact of the Protestant reformations on the Western legal tradition.* Cambridge, MA: Harvard University Press.

Bernhard, N. E. (1993). Ready, willing, able: Network television news and the federal government, 1948–1953. In W. S. Solomon & R. W. McChesney (Eds.), *Ruthless criticism: New perspectives in U. S. communication history* (pp. 291–312). Minneapolis: University of Minnesota Press.

Berninghausen, D. K. (1953). The history of the ALA intellectual freedom committee. *Wilson Library Bulletin, 27*(10), 813–817.

Berry, J. N. (1996). The lessons of SFPL. *Library Journal, 121*(19), 6.

Bertot, J. C., Jaeger, P. T., Langa, L. A., & McClure, C. R. (2006a). Public access computing and Internet access in public libraries: The role of public libraries in e-government and emergency situations. *First Monday, 11*(9). Available: http://www.firstmonday.org/issues/issue11_9/bertot/index.html

———. (2006b). Drafted: I want you to deliver e-government. *Library Journal, 131*(13), 34–39.

Bertot, J. C., Jaeger, P. T., Shuler, J. A., Simmons, S. N., & Grimes, J. M. (2009). Reconciling government documents and e-government: Government information in policy, librarianship, and education. *Government Information Quarterly, 26*, 433–436.

Bertot, J. C., McClure, C. R., & Jaeger, P. T. (2008). Public libraries and the Internet 2007: Issues, implications, and expectations. *Library & Information Science Research, 30*, 175–184.

Bier, V. (2006). Hurricane Katrina as bureaucratic nightmare. In R. J. Daniels, D. F. Kettl, & H. Kunreuther (Eds). *On risk and disaster: Lessons from Hurricane Katrina* (pp. 243–254). Philadelphia: University of Pennsylvania Press.

Bimber, B. (2003). *Information and American democracy: Technology in the evolution of political power.* Cambridge: Cambridge University Press.

Bishop, A. P. (1999). Making digital libraries go: Comparing use across genres. *Proceeding of the 4th annual ACM conference on digital libraries* (pp. 94–103). Berkeley: ACM.

Bishop, A. P., Bazzell, I., Mehra, B., & Smith, C. (2001). Ayfa: Social and digital technologies that reach across the digital divide. *First Monday, 6*(4). Available: http://www.firstmonday.org/issues/issue6_4/bishop/index.html

Bishop, A. P., Neumann, L. J., Star, S. L., Merkel, C., Ignacio, E., & Sandusky, R. J. (2000). Digital libraries: Situating use in changing information infrastructure. *Journal of the American Society for Information Science, 51*, 394–413.

Blood, R. (2000). Weblogs: A history and perspective. Available: http://www.rebeccablood.net/essays/weblog_history.html

Boast, R., Bravo, M., & Srinivasan, R. (2007). Return to Babel: Emergent diversity, digital resources, and local knowledge. *Information Society, 23*, 395–403.

Bob, C. (2005). *Marketing a revolution: Insurgents, media, and international activism.* New York: Cambridge University Press.

Boeder, P. (2005). Habermas' heritage: The future of the public sphere in the network society. *First Monday, 10*(9). Available: http://www.uic.edu/htbin/cgi-wrap/bin/ojs/index.php/fm/article/view/1280/1200

Borgmann, A. (1999). *Holding on to reality: The nature of information at the turn of the millennium.* Chicago: University of Chicago Press.

Borko, H. (1968). Information science: What is it? *American Documentation,* 19(1), 3–5.

Bowker, G. C., & Star, S. L. (1999). *Sorting things out: Classification and its consequences.* Cambridge, MA: Massachusetts Institute of Technology Press.

Bowman, J. (2008). Communities of practice: Web 2.0 principles for service in art libraries. *Art Documentation* 27(1), 4–12.

Bowrey, K., & Rimmer, M. (2005). Rip, mix, burn: The politics of peer to peer and copyright law. *First Monday,* Special Issue 1. Available: http://firstmonday.org/htbin/cgiwrap/bin/ojs/index.php/fm/article/view/1456/1371

Bowring, B. (2008). *The degradation of the international legal order?: The rehabilitation of law and the possibility of politics.* New York: Routledge.

Boyle, J. (1996). *Shamans, software, and spleens: Law and the construction of the information society.* Cambridge, MA: Harvard University Press.

Braman, S. (2004). Where has media policy gone? Defining the field in the 21st century. *Communication Law and Policy,* 9, 153–182.

———. (2006). *Change of state: Information, policy, and power.* Cambridge, MA: Massachusetts Institute of Technology Press.

Brand, A. (1990). *The force of reason: An introduction to Habermas' theory of communicative action.* New York: Allen & Unwin.

Brenkman, J. (2007). *The cultural contradictions of democracy: Political thought since September 11.* Princeton NJ: Princeton University Press.

Brinkley, A. (2008). World War I and the crisis of democracy. In D. Farber (Ed.), *Security v. liberty: Conflicts between civil liberties and national security in American history* (pp. 27–41). New York: Russell Sage Foundation.

Brinkley, D. (2006). *The Great deluge: Hurricane Katrina, New Orleans, and the Mississippi Gulf Coast.* New York: Harper Perennial.

Brito, J., & Dooling, B. (2006, March 25) Who's your daddy? *Wall Street Journal,* A9.

Brophy, P. (2007). *The library in the twenty-first century* (2nd ed.). London: Facet.

Brown, J. S., & Duguid, P. (2002). *The social life of information.* Boston: Harvard Business School Press.

Brown, R. (1985). *Social psychology* (2nd ed.). New York Free Press.

Bruns, A. (2008). Life beyond the public sphere: Towards a networked model for political deliberation. *Information Polity,* 13, 65–79.

Brynin, M., Anderson, B. & Raban, Y. (2007). Introduction. In B. Anderson, M. Brynin, J. Gershung, & Y. Raban (Eds.), *Information and communication technologies in society: E-living in a digital Europe* (pp. 1–17). London: Routledge.

Buchanan, M. (2003). *Nexus: Small worlds and the groundbreaking theory of networks.* New York: Norton.

Buckland, M. (1991a). Information as thing. *Journal of the American Society for Information Science,* 42, 351–360.

———. (1991b). *Information and information systems.* Westport, CT: Praeger.

———. (1997). What is a "document?" *Journal of the American Society for Information Science,* 48, 804–809.

Buckley, S., Duer, K., Mendel, T., & Siochru, S. O. (2008). *Broadcasting, voice, and accountability: A public interest approach to policy, law, and regulation.* Washington DC: World Bank Group.

Budd, J. M. (1995). An epistemological foundation for library and information science. *Library Quarterly,* 65, 295–318.

———. (2003). The library, praxis, and symbolic power. *Library Quarterly,* 73, 19–32.

Burkhart, F. N. (1991). *Media, emergency warnings, and citizen response.* Boulder, CO: Westview.

Burnett, G. (2009). Colliding norms, community, and the place of online information: The case of archive.org. *Library Trends* 57(4), 694–710.

———. (2000). Information exchange in virtual communities: A typology. *Information Research*, 5(4). Available: http://www.shef.ac.uk/~is/publications/infres/paper82.html

———. (2002). The scattered members of an invisible republic: Virtual communities and Paul Ricoeur's Hermeneutics. *Library Quarterly*, 72, 155–178.

Burnett, G., Besant, M., & Chatman, E. A. (2001). Small worlds: Normative behavior in virtual communities and feminist bookselling. *Journal of the American Society for Information Science and Technology*, 52, 536–547.

Burnett, G., & Bonnici, L. (2003). Beyond the FAQ: Implicit and explicit norms in Usenet newsgroups. *Library and Information Science Research*, 25, 333–351.

Burnett, G., & Buerkle, H. (2004). Information exchange in virtual communities: A comparative study. *Journal of Computer-Mediated Communication*, 9(2). Available: http://jcmc.indiana.edu/vol9/issue2/burnett.html

Burnett, G., & Jaeger, P. T. (2005). Recent policy developments and international students in the United States: Implications for information access and exchange. Paper presented at the 2005 *Association for Library & Information Science Education Conference*.

———. (2008). Small worlds, lifeworlds, and information: The ramifications of the information behaviors of social groups in public policy and the public sphere. *Information Research*, 13(2), paper 346. Available: http://InformationR.net/ir/13-2/paper346.html

Burnett, G., Jaeger, P. T., & Thompson, K. M. (2008). The social aspects of information access: The viewpoint of normative theory of information behavior. *Library & Information Science Research*, 30, 56–66.

Buschman, J. E. (2003). *Dismantling the public sphere: Situating and sustaining librarianship in the age of the new public philosophy*. Westport, CT: Libraries Unlimited.

———. (2006). "The integrity and obstinacy of intellectual creations": Jürgen Habermas and librarianship's theoretical literature. *Library Quarterly*, 76, 270–299.

———. (2007a). Democratic theory in library and information science: Toward an emendation. *Journal of the American Society for Information Science and Technology*, 58, 1483–1496.

———. (2007b). Transgression or statis? Challenging Foucault in LIS theory. *Library Quarterly*, 77, 21–44.

Buschman, J. E., & Leckie, G. J. (Eds.). (2007). *The library as place: History, community, and culture*. Westport, CT: Libraries Unlimited.

Butler, R. P. (2003). Copyright law and organizing the Internet. *Library Trends*, 52(2), 307–317.

Caidi, N., & Allard, D. (2005). Social inclusion of newcomers to Canada: An information problem? *Library & Information Science Research*, 27, 302–324.

Caidi, N., & Ross, A. (2005). Information rights and national security. *Government Information Quarterly*, 22, 663–684.

Carlson, S. (2005). Whose work is it, anyway? *Chronicle of Higher Education*, 51(47), A33–A35.

Carlton, J. (2009, January 19). Folks are flocking to the library, a cozy place to look for a job: Books, computers and wi-fi are free, but staffs are stressed by crowds, cutbacks. *Washington Post*, A1.

Carr, N. (2008, July). Is Google making us stupid? *Atlantic*. Available: http://www.theatlantic.com/doc/200807/google

Carrico, J. C., & Smalldon, K. L. (2004). Licensed to ILL: A beginning guide to negotiating e-resources licenses to permit resource sharing. *Journal of Library Administration*, 40(1–2), 41–54.

Cary, K., & Ogburn, J. L. (2000). Developing a consortial approach to catalog-ing and intellectual access. *Library Collections, Acquisitions, & Technical Services, 24,* 45–51.

Case, D. O. (2002). *Looking for information: A survey of research on information seeking, needs, and behavior.* San Diego, CA: Academic Press.

Castells, M. (2000). *The information age: Economy, society and culture, volume 1, the rise of the network society.* Oxford: Blackwell Publishers.

Cavelty, M. D. (2008). *Cyber-security and threat politics: US efforts to secure the information age.* London: Routledge.

Chadwick, A. (2001). The electronic face of government in the Internet age: Borrowing from Murray Edelman. *Information, Communication & Society, 4*(3), 435–457.

Chakravartty, P., & Zhao, Y. (2008). *Global communications: Toward a transcultural political economy.* Lanham, MD: Rowman & Littlefield.

Chatman, E. A. (1987). The information world of low-skilled workers. *Library & Information Science Research, 9,* 265–283.

———. (1991a). Channels to a larger social world: Older women staying in contact with the great society. *Library & Information Science Research, 9,* 281–300.

———. (1991b). Life in a small world: Applicability of gratification theory to infor-mation-seeking behavior. *Journal of the American Society for Information Science, 42,* 438–449.

———. (1992). *The information world of retired women.* Westport, CT: Green-wood.

———. (1999). A theory of life in the round. *Journal of the American Society for Information Science, 50,* 207–217.

———. (2000). Framing social life in theory and research. *The New Review of Information Behaviour Research, 1,* 3–17.

Chatman, E. A., & Pendleton, V. E. M. (1995). Knowledge gap, information-seek-ing and the poor. *The Reference Librarian, 49–50,* 135–145.

Chen, H., & Rasmussen, E. M. (1999). Intellectual access to images. *Library Trends, 48*(2), 291–302.

Chernilo, D. (2007). A quest for universalism: Re-assessing the nature of classical social theory's cosmopolitanism. *European Journal of Social Theory, 10*(1), 8–17.

Chertoff, M. (2005). *Statement before the Senate Committee on Homeland Secu-rity and Governmental Affairs.* Department of Homeland Security: Second Stage Review, July 14.

Chester, E. W. (1969). *Radio, television, & American politics.* New York: Sheed and Ward.

Choney, S. (2009, March 26). Federal government: Star of YouTube, Flickr? *MSNBC.* Available: http://www.msnbc.msn.com/id/29882992/

Ciffolilli, A. (2003). Phantom authority, self-selective recruitment and retention of members in virtual communities: The case of Wikipedia. *First Monday, 8*(12). Available: http://www.firstmonday.org/issues/issue8_12/ciffolilli/index.html

Clark, W. (2000). *Activism in the public sphere: Exploring the discourse of politi-cal participation.* Burlington, CT: Ashgate.

Clarke, R. (2001). "Information wants to be free. . . ." Available: http://www.rog-erclarke.com/II/IWtbF.html

Clayman, S. E. (2004). Arenas of interaction in the mediated public sphere. *Poet-ics, 32,* 29–49.

Cleaver, H. (1998). The Zapatistas and the electronic fabric of struggle. In J. Hollo-way & E. Peleaz (Eds.), *Zapatista! Reinventing revolution in Mexico.* London: Pluto Press.

Cohen, N. (2009, June 12). Twitter on the barricades: Six lessons learned. *New York Times, Week in Review,* 4.

Coleman, J. (1988). Social capital in the creation of human capital. *American Jour-nal of Sociology, 94,* 95–120.

Comaromi, J. P. (1990). Summation of classification as an enhancement of intellectual access to information in an online environment. *Cataloging & Classification Quarterly, 11*(1), 99–102.

Comellas, F., & Sampels, M. (2002). Deterministic small–world networks. *Physica A, 309*, 231–235.

Comfort, L. K. (2007). Crisis management in hindsight: Cognition, communication, coordination, and control. *Public Administration Review, 67*(special issue), 189–197.

Committee on Government Reform. (2004). *Secrecy in the Bush administration.* Washington, DC: Author.

Conant, R. W. (1965). *The public library and the city.* Cambridge, MA: MIT.

Conquest, R. (2000). *Reflections on a ravaged century.* New York: Norton.

Cooper, C., & Block, R. (2006). *Disaster: Hurricane Katrina and the failure of homeland security.* New York: Times Books.

Cornelius, I. (1996). *Meaning and method in information studies.* Norwood, NJ: Ablex.

Corner, J. (1995). *Television form and public address.* London: Edward Arnold.

Cossette, A. (1976/2009). *Humanism and libraries: An essay on the philosophy of librarianship* (R. Litwin, Trans.). Duluth, MN: Library Juice Press.

Courtney, N. (2007). *Library 2.0 and beyond: Innovative technology and tomorrow's user.* Greenwood, CT: Libraries Unlimited.

Covello, V. T. (1996). Communicating risk is crisis and noncrisis situations: Tools and techniques for effective environmental communication. In R. Kolluru, S. Bartell, R. Pitblado, & S. Stricoff, (Eds.). *Risk assessment and management handbook for environmental, health, and safety professionals* (pp. 15.3–15.15). New York: McGraw Hill.

Crumlish, C. (2004). *The power of many: How the living web is transforming politics, business, and everyday life.* San Francisco: Sybex.

Cuadra, C. A. (Ed.) (1966). *Annual review of information science and technology, volume 1.* New York: John Wiley & Sons.

Culnan, M. J. (1983). Environmental scanning: The effects of task complexity and source accessibility on information gathering behavior. *Decision Sciences, 14*(2), 194–206.

———. (1984). The dimensions of accessibility to online information: Implications for implanting office information systems. *ACM Transactions on Office Information Systems, 2*(2), 141–150.

———. (1985). The dimensions of perceived accessibility to information: Implications for the delivery of information systems and services. *Journal of the American Society for Information Science, 36*(5), 302–308.

Dahl, R. A. (1961). *Who governs? Democracy and power in an American city.* New Haven, CT: Yale University Press.

———. (1982). *Dilemmas of pluralist democracy: Autonomy vs. control.* New Haven, CT: Yale University Press.

———. (1989). *Democracy and its critics.* New Haven, CT: Yale University Press.

———. (1994). The new American political (dis)order. In R. Dahl (Ed.), *The new American political (dis)order.* Berkeley: Institute of Governmental Studies.

Dahlgren, P. (1995). *Television and the public sphere: Citizenship, democracy and the media.* New York: Sage.

Dandeker, C. (1990). *Surveillance, power and modernity: Bureaucracy and discipline from 1700 to the present day.* New York: St. Martin's.

Daniel, M. (2009). *Scandal & civility: Journalism and the birth of American democracy.* Oxford: Oxford University Press.

Dann, G.E., & Haddow, N. (2008). Just doing business, or doing just business: Google, Microsoft, Yahoo!, and the business of censoring China's Internet. *Journal of Business Ethics, 79*, 219–234.

Darnton, R. (1995). *Forbidden bestsellers of pre-revolutionary France*. New York: Norton.

Davies, D. W. (1974). *Public libraries as culture and social centers: The origin of the concept*. Metuchen, NJ: Scarecrow.

Davies, N. (1999). *The Isles: A history*. New York: Oxford University Press.

Davis, R. (1998). *The Web of politics*. London: Oxford University Press.

Davis, R., & Owen, D. (1998). *New media in American politics*. London: Oxford University Press.

Dearstyne, B. W. (2005). Fighting terrorism, making war: Critical insights in the management of information and intelligence. *Government and Information Quarterly, 22*, 170–186.

———. (2006). Taking charge: Disaster fallout reinforces RIM's importance. *The Information Management Journal*, July/August, 37–42.

———. (2007). The FDNY on 9/11: Information and decision making in crisis. *Government and Information Quarterly, 24*, 29–46.

Deibert, R., & Rohozinski, R. (2008). Good for liberty, bad for security? Global civil society and the securitization of the Internet. In R. Deibert, J. Palfrey, R. Rohozinski, & J. Zittrain (Eds.), *Access denied: The practice and policy of global Internet filtering* (pp. 123–150). Cambridge, MA: Massachusetts Institute of Technology Press.

Deines-Jones, C. (1996). Access to library Internet services for patrons with disabilities: Pragmatic considerations for developers. *Library Hi Tech, 14*(1), 57–64.

DeMott, J. (1990). "Company line" raises ethical dilemma. *Media Law Notes* (Spring), 9.

Dervin, B. (1973). Information needs of urban residents: A conceptual context. In E. S. Warner, A. D. Murray, & V. E. Palmour (Eds.), *Information needs of urban residents* (pp. 8–42). Washington, DC: Department of Health, Education, and Welfare.

———. (1994). Information—democracy: An examination of underlying assumptions. *Journal of the American Society for Information Science, 45*, 369–385.

Dervin, B., & Nilan, M. (1986). Information needs and uses. In M. E. Williams (Ed.), *Annual review of information science and technology* (Vol. 21, pp. 1–25). White Plains, NY: Knowledge Industry.

Dewey, J. (1959). *Dialogue on John Dewey*. New York: Horizon.

Diaz, A. (2008). Through the Google goggles: Sociopolitical bias in search engine design. In A. Spink, & M. Zimmer (Eds.), *Web search: Multidisciplinary perspectives* (pp. 13–34). Berlin: Springer-Verlag.

Dibble, J. (1998). *My tiny life: Crime and passion in a virtual world*. New York: Henry Holt and Company.

Dilevko, J., & Dali, K. (2003). Electronic databases for readers' advisory services and intellectual access to translated fiction not originally written in English. *Library Resources & Technical Services, 47*(3), 80–95.

Dillon, A., & Norris, A. (2005). Crying wolf: An examination and reconsideration of the perception of crisis in LIS education. *Journal of Education for Library and Information Science, 46*(3), 280–298.

DiMaggio, P. (1997). Culture and cognition. *Annual Review of Sociology, 23*, 263–287.

DiMaggio, P., Evans, J., & Bryson, B. (1997). Have Americans' social attitudes become more polarized? In R. H. Williams (Ed.), *Cultural wars in American politics: Critical reviews of a popular myth* (pp. 63–100). New York: Aldine De Gruyter.

Ditzion, S. H. (1947). *Arsenals of a democratic culture*. Chicago: American Library Association.

Doctorow, C. (2006). Giving it away. *Forbes.com*. Available: http://www. forbes.com/2006/11/30/cory-doctorow-copyright-tech-media_cz_cd_ books06_1201doctorow.html

Dorogovtsev, S. N. and Mendes, J. F. F. (2003). *Evolution of networks: From biological networks to the Internet and WWW*. New York: Oxford University Press.

Dowell, D. R. (2008). The "i" in libraries. *American Libraries, 39*(1/2), 42.

Drabenstott, K. M., & Atkins, D. E. (1996). The Kellogg CRISTAL-Ed project: Creating a model program to support libraries in the digital age. In I. P. Godden (Ed.), *Advances in librarianship* (20th ed.). San Diego, CA: Academic Press.

Dresang, E. T. (2006). Intellectual freedom and libraries: Complexity and change in the twenty-first century digital environment. *Library Quarterly, 76*, 169–192.

Duff, A. S. (2000). *Information society studies*. London: Routledge.

Duffy, R., & Everton, R. (2008). Media, democracy, and the state in Venezuela's "Bolivarian revolution." In P. Chakravartty & Y. Zhao (Eds.), *Global communications: Toward a transcultural political economy* (pp. 113–140). Lanham, MD: Rowman & Littlefield.

Duguid, P. (1996). Material matters: The past and futurology of the book. In G. Nunberg (Ed.), *The future of the book* (pp. 63–102). Berkeley: University of California Press.

DuMont, R. R. (1977). *Reform and reaction: The big city public library in American life*. Westport, CT: Greenwood.

Durrance, J. C., & Fisher, K. E. (2002). *Online community information: Creating a nexus at your library*. Chicago: American Library Association.

Dyson, M. E. (2005). *Come hell or high water: Hurricane Katrina and the color of disaster*. New York: Basic Books.

Eberhard, W. (2000). The threat from within: Balancing access and national security. In C. N. Davis & S. L. Splichal (Eds.), *Access denied: Freedom of information in the information age*. Ames: Iowa State University Press.

Eilperin, J. (2006). *Fight club politics: How partisanship is poisoning the House of Representatives*. New York: Rowman & Littlefield.

El-Abbadi, M. (1990). *The life and fate of the ancient library of Alexandria*. Paris: UNESCO.

Ellis, J. J. (1996). *American sphinx: The character of Thomas Jefferson*. New York: Vintage.

Elmer-DeWitt, P., Jackson, D.S., & King, W. (1993). First nation in cyberspace. *Time*. Available: http://www.time.com/time/magazine/article/0,9171,979768,00. html

Emerson, T. I. (1984). National security and civil liberties: Introduction. *Cornell Law Review, 69*, 685–689.

Etzioni, A. (1993). *The spirit of community: Rights, responsibilities, and the communitarian agenda*. New York: Crown Publishers.

Ewen, S. (1996). *PR!: The social history of spin*. New York: Basic.

Feinberg, L. E. (2004). FOIA, Federal information policy, and information availability in a post-9/11 world. *Government Information Quarterly, 21*, 439–460.

Ferling, J. (2000). *Setting the world ablaze: Washington, Adams, Jefferson, and the American Revolution*. New York: Oxford University Press.

Ferullo, D. L. (2004). Major copyright issues in academic libraries: Legal implications of a digital environment. *Journal of Library Administration, 40*(1–2), 23–40.

Fidel, R. & Green, M. (2004). The many faces of accessibility: Engineers' perception of information sources. *Information Processing & Management, 40*(3), 563–581.

Fisher, K. E., Durrance, J. C., & Hinton, M. B. (2004). Information grounds and the use of need-based services by immigrants in Queens, New York: A context-based, outcome evaluation approach. *Journal of the American Society of Information Science and Technology, 55,* 754–766.

Fisher, K. E., Erdelez, S., McKechnie, L. E. F. (Eds). (2005). *Theories of information behavior.* Medford, NJ: Information Today.

Fisher, K. E., & Naumer, C. M. (2005). Information grounds: Theoretical basis and empirical findings on information flow in social settings. In A. Spink & C. Cole (Eds.), *New directions in human information behavior* (pp. 93–111). New York: Springer.

Fisher, K. E., Naumer, C. M., Durrance, J. C., Stromski, L., & Christiansen, T. (2005). Something old, something new: Preliminary findings from an exploratory study about people' information habits and information grounds. *Information Research, 10*(2), paper 223. Available: http://InformationR.net/ir/10–2/paper223.html

Fishkin, J. S. (1991). *Democracy and deliberation: New direction for democratic reform.* New Haven, CT: Yale University Press.

Fiske, M. (1959). *Book selection and censorship: A study of school and public libraries in California.* Berkeley: University of California Press.

Foerstel, H. N. (1991). *Surveillance in the stacks: The FBI's library awareness program.* Westport, CT: Greenwood.

———. (2004). *Refuge of a scoundrel: The Patriot Act in libraries.* Westport, CT: Libraries Unlimited.

Foucault, M. (1979). *Discipline and punishment: The birth of the prison.* New York: Penguin.

Fourie, D. K., & Dowell, D. R. (2002). *Libraries in the information age: An introduction and career exploration.* Westport, CT: Libraries Unlimited.

Freedman, D. (2008). *The politics of media policy.* Cambridge: Polity.

Freedom of Information Act, 5 U.S.C. secs. 552 et seq.

Friedland, R., & Alford, R. R. (1991). Bringing society back in: Symbols, practices, and institutional contradictions. In W. W. Powell & P. J. DiMaggio (eds.), *The new institutionalism in organizational analysis* (pp. 232–266). Chicago: University of Chicago Press.

Friedman, T. (1999). The semiotics of SimCity. *First Monday,* 4(4–5). Available: http://firstmonday.org/htbin/cgiwrap/bin/ojs/index.php/fm/article/view/660/575

Fritz, B., Keefer, B., & Nyhan, B. (2004). *All the president's spin: George W. Bush, the media, and the truth.* New York: Touchstone.

Fry, J. (2006). Google's privacy responsibilities at home and abroad. *Journal of Librarianship and Information Science, 38*(3), 135–138.

Fuchs, C. (2008). *Internet and society: Social theory in the information age.* New York: Routledge.

Garnett, J. L., & Kouzmin, A. (2007). Communicating through Katrina: Competing and complementary conceptual lenses on crisis communication. *Public Administration Review,* 67(special issue), 171–188.

Garrison, D. (1993). *Apostles of culture: The public librarian and American society, 1876–1920.* Madison: University of Wisconsin Press.

Gasaway, L. N. (2000). Values conflict in the digital environment: Librarians versus copyright holders. *Columbia—VLA Journal of Law & the Arts,* Fall, 115–161.

Gathegi, J. N. (2005). The public library and the (de)evolution of a legal doctrine. *Library Quarterly, 75,* 1–19.

Gellar, E. (1974). Intellectual freedom: Eternal principle or unanticipated consequence? *Library Journal, 99,* 1364–1367.

———. (1984). *Forbidden books in American public libraries, 1876–1939: A study in cultural change.* Westport, CT: Greenwood.

Gellman, B., & Becker, J. (2007a, June 24). "A different understanding with the president." *Washington Post.* Available: http://blog.washingtonpost.com/cheney/chapters/chapter_1/

———. (2007b, June 25). Pushing the envelope on presidential power. *Washington Post.* Available: http://blog.washingtonpost.com/cheney/chapters/pushing_the_envelope_on_presi/index.html

Gerard, D. (1978). *Libraries in society: A reader.* London: Clive Bingley.

Giacomello, G. (2005). *National governments and the control of the Internet: A digital challenge.* London: Routledge.

Giddens, A. (1985). *The nation-state and violence.* Cambridge: Polity Press.

———. (1991). Structuration theory: Past, present, and future. In C. G. S. Bryant & D. Jary (Eds.), *Giddens' theory of structuration: A critical appreciation* (pp. 202–221). London: Routledge.

Giles, J. (2005). Internet encyclopedias go head to head. *News@Nature.com.* Available: http://www.nature.com/news/2005/051212/full/438900a.html

Gilliland, A. J. (1988). Introduction: Automating intellectual access to archives. *Library Trends, 38,* 495–499.

Gilman-Opalsky, R. (2008). *Unbounded publics: Transgressive public spheres, Zapatismo, and political theory.* Lanham, MD: Lexington.

Gitelman, L. (1999). *Scripts, grooves, and writing machines: Representing technology in the Edison era.* Stanford, CA: Stanford University Press.

———. (2006). *Always already new: Media, history, and the data of culture.* Cambridge, MA: Massachusetts Institute of Technology Press.

Given, L., & Leckie, G. L. (2003). Sweeping the library: Mapping the social activity space of the public library. *Library and Information Science Research, 25,* 365–385.

Goldwin, R. A. (1986). Of men and angels: A search for morality in the Constitution. In R. H. Horowitz (Ed.), *The moral foundations of the American republic* (3rd ed.) (pp. 42–61). Charlottesville: University of Virginia.

Gorham-Oscilowski, U., & Jaeger, P. T. (2008). National Security Letters, the USA PATRIOT Act, and the Constitution: The tensions between national security and civil rights. *Government Information Quarterly, 25*(4), 625–644.

Gorman, M. (2004). What ails library education? *Journal of Academic Librarianship, 30*(2), 99–101.

Gough, S., & Stables, A. (2008). *Sustainability and security within liberal societies.* London: Routledge.

Goulding, A. (2001). Information poverty or overload? *Journal of Librarianship and Information Science, 33,* 109–111.

———. (2004). Libraries and social capital. *Journal of Librarianship and Information Science, 36*(1), 3–6.

Granovetter, M. (1973). The strength of weak ties. *American Journal of Sociology, 78,* 1360–1380.

Gray, C. M. (1993). The civic role of libraries. In J. E. Buschman (Ed.), *Critical approaches to information technology in librarianship: Foundations and applications.* Westport, CT: Greenwood.

Green, J. (2007). Subscription libraries and commercial circulating libraries in colonial Philadelphia and New York. In T. Augst & K. Carpenter (Eds.), *Institutions of reading: The social life of libraries in the United States* (pp. 24–52). Amherst: University of Massachusetts Press.

Green, L. (2001). *Communication, technology and society.* Sage: London.

Greenslade, R. (2003, February 17). Their masters' voice. *The Guardian.* Available: http://www.guardian.co.uk/media/2003/feb/17/mondaymediasection.iraq

Grimes, J. M., Jaeger, P. T., & Fleischmann, K. R. (2008). Obfuscatocracy: Contractual frameworks in the governance of virtual worlds. *First Monday, 13*(9). Available: http://www.uic.edu/htbin/cgiwrap/bin/ojs/index.php/fm/article/view/2153/2029

Grusin, R. (2004). Pre-meditation. *Criticism, 46*(1), 17–39.

Gup, T. (2007). *Nation of secrets: The threat to democracy and the American way of life.* New York: Doubleday.

Habermas, J. (1981). *The structural transformation of the public sphere: An inquiry into category of bourgeois society.* Cambridge, MA: Polity.

———. (1984). *The theory of communicative action, volume 1, reason and rationalization of society.* Cambridge, MA: Polity.

———. (1989). *The structural transformations of the public sphere: An inquiry into a category of bourgeois society.* Cambridge, MA: Massachusetts Institute of Technology Press.

———. (1992). Further reflections on the public sphere. In J. Calhoun (Ed.), *Critical social theory: Culture, theory and the challenge of difference* (pp. 421–462). Oxford: Blackwell.

———. (1996a). *Between facts and norms: Contributions to a discourse theory of law and democracy.* Cambridge, MA: Polity.

———. (1996b). The normative content of modernity. In W. Outhewiate (Ed.), *The Habermas reader* (pp. 341–365). Cambridge, MA: Polity.

———. (2001). *The postnational constellation: Political essays.* Cambridge, MA: Polity.

Hacker, J. S., & Pierson, P. (2005a). Abandoning the middle: The Bush tax cuts and the limits of democratic control. *Perspectives on Politics, 3*(1), 33–53.

———. (2005b). *Off-center: The republican revolution and the erosion of American democracy.* New Haven, CT: Yale.

———. (2007). Tax politics and the struggle over activist government. In P. Pierson, & T. Skocpol (Eds.), *The transformation of the American politics: Activist government and the rise conservatism* (pp. 256–280). New York: Russell Sage.

Hacker, K. L. (1996). Missing links in the evolution of electronic democratization. *Media, Culture & Society, 18,* 213–232.

Haddow, G. D., & Bullock, J. A. (2003). *Introduction to emergency management.* Newton, PA: Buttersworth-Heinemann.

Haiman, F. S. (1981). *Speech and law in a free society.* Chicago: University of Chicago Press.

Halstead, T. J. (2008). Presidential signing statements: Executive aggrandizement, judicial ambivalence and congressional vituperation. *Government Information Quarterly, 25,* 563–591.

Hamburger, T., & Wallsten, P. (2006). *One party country: The republican plan for dominance in the 21st century.* Hoboken, NJ: Wiley & Sons.

Han, S. (2008). *Navigating technomedia: Caught in the Web.* Lanham MD: Rowman & Littlefield.

Hanratty, E. (2005). Google library: Beyond fair use? *Duke Law & Technology Review, 10.* Available: http://www.law.duke.edu/journals/dltr/articles/pdf/2005dltr0010.pdf

Hanson, E. C. (2008). *The information revolution and world politics.* Lanham, MD: Rowman & Littlefield.

Hansson, J. (2005). Hermeneutics as bridge between the modern and the postmodern in library and information science. *Journal of Documentation, 61,* 102–113.

Harris, M. H. (1973). The purpose of the American public library: A revisionist interpretation of history. *Library Journal, 98,* 2509–2514.

———. (1976). Public libraries and the decline of the democratic dogma. *Library Journal, 101,* 2225–2230.

———. (1986). The dialectic of defeat: Antimonies in research in library and information science. *Library Trends, 34,* 515–531.

Hartman, T. (2007). The changing definition of U.S. libraries. *Libri, 57,* 1–8.

Hauptman, R. (2002). *Ethics and librarianship*. Jefferson, NC: McFarland & Company.

Heckart, R. J. (1991). The library as marketplace of ideas. *College and Research Libraries, 52*, 491–505.

Heiskanen, E. (2002). The institutional logic of life cycle thinking. *Journal of Cleaner Production, 10*, 427–437.

Helft, M. (2009, May 4). Libraries ask judge to monitor Google Books settlement. *New York Times*, A1, B4.

Heres, J., & Thomas, F. (2007). Civic participation and ICTs. In B. Anderson, M. Brynin, J. Gershung, & Y. Raban (Eds.), *Information and communication technologies in society: E-living in a digital Europe* (pp. 175–188). London: Routledge.

Herman, E. S., & McChesney, R. W. (1997). *The global media: The new visionaries of global capitalism*. London: Cassell.

Hernon, P., Relyea, H. C., Dugan, R. E., & Cheverie, J. F. (2002). *United States government information: Policies and sources*. Westport, CT: Libraries Unlimited.

Hernon, P., & Schwartz, C. (2009). Research by default. *Library & Information Science Research, 31*, 137.

Herring, S. C., Kouper, I., Scheidt, L. A., & Wright, E. L. (2004). Women and children last: The discursive construction of weblogs. *Into the blogosphere: Rhetoric, community, and culture of weblogs*. Available: http://blog.lib.umn.edu/blogosphere/women_and_children.html

Herring, S. C., Scheidt, L. A., Bonus, S., & Wright, E. (2004). Bridging the gap: A genre analysis of weblogs. *Proceedings of the Thirty-seventh Hawaii International Conference on System Sciences (HICSS-37)*. Los Alamitos: IEEE Press.

Hersberger, J. (2002). Are the economically poor information poor? Does the digital divide affect the homeless and information access? *Canadian Journal of Information and Library Science, 27*(3), 45–63.

———. (2003). A qualitative approach to examining information transfer via social networks among homeless populations. *New Review of Information Behaviour Research, 4*, 63–78.

Hess, A. (2008). Reconsidering the rhizome: A textual analysis of web search engines as gatekeepers of the internet. In A. Spink & M. Zimmer (Eds.), *Web search: Multidisciplinary perspectives* (pp. 35–50). Berlin: Springer-Verlag.

Hiebert, R. E. (2003). Public relations and propaganda in framing the Iraq war: a preliminary review. *Public Relations Review, 29*, 243–255.

———. (2005). Commentary: new technologies, public relations, and democracy. *Public Relations Review, 31*, 1–9.

Hindman, M. (2008). What is the online public sphere good for? In J. Turow & L. Tsui (Eds.), *The hyperlinked society: Questioning connections in the digital age* (pp. 268–288). Ann Arbor: University of Michigan Press.

Hitchens, L. (2006). *Broadcasting pluralism and diversity: A comparative study of policy and regulation*. Portland, OR: Hart.

Hobson, C. F. (1996). *The great Chief Justice: John Marshall and the rule of law*. Lawrence: University Press of Kansas.

Hofstetter, C. R., Barker, D., Smith, J. T., Zari, G. M., & Ingrassia, T. A. (1999). Information, misinformation, and political talk radio. *Political Research Quarterly, 52*(2), 353–369.

Hogenboom, K. (2008). Lessons learned about access to government information after World War II can be applied after September 11. *Government Information Quarterly, 25*, 90–103.

Homeland Security Act, Public Law 107–295.

Hosein, I. (2004). The sources of law: Policy dynamics in a digital and terrorized world. *Information Society, 20*, 187–199.

Houser, L. J., & Schrader, A. M. (1978). *The search for a scientific profession: Library science education in the U.S. and Canada.* Metuchen, NJ: Scarecrow.

Howard, M. (1983). *The causes of war.* London: Unwin Counterpoint.

Hurlbert, J. M., Savidge, C. R., & Laudenslager, G. R. (2003). Process-based assignments: How promoting information literacy prevents plagiarism. *College & Undergraduate Libraries, 10*(1), 39–51.

Husserl, E. (1962). *Ideas: General introduction to pure phenomenology.* New York: Collier.

———. (1970). *The crisis of the European sciences and transcendental phenomenology.* New York: Kluwer.

Intner, S. S. (1991). Intellectual access to patron-use software. *Library Trends, 40*(1), 42–62.

Jackson, S. L. (1974). *Libraries and librarianship in the West: A brief history.* New York: McGraw-Hill.

Jacobs, A. (2009a, February 5). Chinese learn limits of online freedom as the filter tightens. *New York Times*, A8. Available: http://www.nytimes.com/2009/02/05/world/asia/05beijing.html?_r=1

———. (2009b, June 11). China faces criticism over new software censor. *New York Times*, A12. Available: http://www.nytimes.com/2009/06/11/world/asia/11censor.html?_r=1&scp=1&sq=China%20software%20filters&st=cse

Jacobs, L. R., & Shaprio, R. (2000). *Politicians don't pander: Political manipulation and the loss of democratic responsiveness.* Chicago: University of Chicago Press.

Jaeger, P. T. (2005). Deliberative democracy and the conceptual foundations of electronic government. *Government Information Quarterly, 22*(4), 702–719.

———. (2007). Information policy, information access, and democratic participation: The national and international implications of the Bush administration's information politics. *Government Information Quarterly, 24*, 840–859.

———. (2009a). Public libraries and local e-government. In C. G. Reddick (Ed.), *Handbook on research on strategies for local e-government adoption and implementation: Comparative studies* (pp. 647-660). Hershey, PA: IGI Global.

———. (2009b). The fourth branch of government and the historical legacy of the Bush administration's information policies. *Government Information Quarterly, 26*, 311–313.

Jaeger, P. T., Bertot, J. C., & McClure, C. R. (2003). The impact of the USA Patriot Act on collection and analysis of personal information under the Foreign Intelligence Surveillance Act. *Government Information Quarterly, 20*(3), 295–314.

Jaeger, P. T., & Bowman, C. A. (2005). *Understanding disability: Inclusion, access, diversity, & civil rights.* Westport, CT: Praeger.

Jaeger, P. T., & Burnett, G. (2003). Curtailing online education in the name of homeland security: The USA PATRIOT Act, SEVIS, and international students in the United States. *First Monday, 8*(9). Available: http://firstmonday.org/issues/issue8_9/jaeger/index.html

———. (2005). Information access and exchange among small worlds in a democratic society: The role of policy in redefining information behavior in the post-9/11 United States. *Library Quarterly, 75*(4), 464–495.

Jaeger, P. T., & Fleischmann, K. R. (2007). Public libraries, values, trust, and e-government. *Information Technology and Libraries, 26*(4), 35–43.

Jaeger, P. T., Fleischmann, K. R., Preece, J., Shneiderman, B., Wu, F. P., & Qu, Y. (2007). Community response grids: Facilitating community response to biosecurity and bioterror emergencies through information and communication technologies. *Biosecurity and Bioterrorism, 5*(4), 335–346.

Jaeger, P. T., Langa, L. A., McClure, C. R., & Bertot, J. C. (2006). The 2004 and 2005 Gulf Coast hurricanes: Evolving roles and lessons learned for public libraries in disaster preparedness and community services. *Public Library Quarterly, 25*(3/4), 199–214.

Jaeger, P. T., Lin, J., & Grimes, J. (2008). Cloud computing and information policy: Computing in a policy cloud? *Journal of Information Technology & Politics, 5*(3), 269–283.

Jaeger, P. T., Lin, J., Grimes, J. M., & Simmons, S. N. (2009). Where is the cloud? Geography, economics, environment, and jurisdiction in cloud computing. *First Monday, 14*(5). Available: http://www.uic.edu/htbin/cgiwrap/bin/ojs/index.php/fm/article/view/2456/2171

Jaeger, P. T., & McClure, C. R. (2004). Potential legal challenges to the application of the Children's Internet Protection Act (CIPA) in public libraries: Strategies and issues. *First Monday, 9*(2). Available: http://www.firstmonday.org/issues/issue9_2/jaeger/index.html.

Jaeger, P. T., McClure, C. R., Bertot, J. C., & Snead, J. T. (2004). The USA PATRIOT Act, the Foreign Intelligence Surveillance Act, and information policy research in libraries: Issues, impacts, and questions for library researchers. *Library Quarterly, 74*(2), 99–121.

Jaeger, P. T., Shneiderman, B., Fleischmann, K. R., Preece, J., Qu, Y., & Wu, F. P. (2007). Community response grids: E-government, social networks, and effective emergency response. *Telecommunications Policy, 31,* 592–604.

Jaeger, P. T., & Yan, Z. (2009). One law with two outcomes: Comparing the implementation of the Children's Internet Protection Act in public libraries and public schools. *Information Technology and Libraries, 28*(1), 8–16.

Jerome, F. (2002). *The Einstein file: J. Edgar Hoover's secret war against the world's most famous scientist.* New York: St. Martin's.

Johnson, B. B., & Slovic, P. (1995). Presenting uncertainty in health risk assessment: Initial studies of its effects on risk perception and trust. *Risk Analysis, 15,* 485–494.

Johnson, C., & Warrick, J. (2009, March 3). CIA destroyed 92 interrogation tapes, probe says. *Washington Post,* A1, A5.

Johnson, T. J., & Kaye, B. K. (2004). Wag the blog: How reliance on traditional media and the internet influence credibility perceptions of weblogs among blog users. *J&MC Quarterly, 81*(3), 622–642.

Jones, C., & Mitnick, S. (2006). Open source disaster recovery: Case studies of networked collaboration. *First Monday, 11*(5). Available: http://www.firstmonday.org/issues/issue11_5/jones/

Jones, P. A., Jr. (1993). From censorship to intellectual freedom to empowerment: The evolution of the social responsibility of the American public library. *North Carolina Libraries, 52,* 135–137.

Jul, S., & Furnas, G. W. (1998). Critical zones in desert fog: Aids to multiscale navigation. *Proceeding of the 11th annual ACM symposium on user interface software and technology* (pp. 97–106). San Francisco: ACM.

Julien, H., McKechnie, L. E. F., & Hart, S. (2005). Affective issues in library and information science systems work: A content analysis. *Library & Information Science Research, 27,* 453–466.

Jung, C. (2003). The politics of indigenous identity: Neoliberalism, cultural rights, and the Mexican Zapatistas. *Social Research, 70*(2), 433–461.

Kakabadse, A., Kakebadse, N. K., & Kouzmin, A. (2003). Reinventing the democratic governance project through democracy? A growing agenda for debate. *Public Administration Review, 63*(1), 44–60.

Kapucu, N. (2004). Interagency communication networks during emergencies: Boundary spanners in multiagency coordination. *American Review of Public Administration, 36,* 207–225.

Katz, E. (1996). And deliver us from segmentation. *The Annals of the American Academy of Political Scientists, 546,* 22–33.

Kelton, K., Fleischmann, K. R., & Wallace, W. A. (2008). Trust in digital information. *Journal of the American Society for Information Science and Technology, 59,* 363–374.

Kermeny, J. G. (1962). A library for 2000 A.D. In M. Greenberger (Ed.), *Computers and the world of the future.* Cambridge, MA: Massachusetts Institute of Technology Press.

Kettl, D. F. (2004). *System under stress: Homeland security and American politics.* Washington, DC: CQ.

Kilworth, P. D., & Bernard, H. R. (1979). A pseudomodel of the small world problem. *Social Forces, 58,* 477–505.

Kim, E., Lee, B., & Menon, N. M. (2008). Social welfare implications of the digital divide. *Government Information Quarterly, 26,* 377–386.

King, J. L., & Lyytinen, K. (2004). Grasp and reach. *Management Information Systems Quarterly, 28*(4), 539–551.

Kitaro, N. (1987). *Last writings: Nothingness and the religious worldview* (D. A. Dilworth, trans). Manoa: University of Hawaii Press.

Klinenberg, E. (2007). *Fighting for air: The battle to control America's media.* New York: Metropolitan.

Klotz, R. J. (2004). *The politics of Internet communication.* Lanham, MD: Rowman & Littlefield.

Knezo, G. J. (2003). *Sensitive but unclassified and other federal security controls on scientific and technical information: History and current controversy.* Washington, DC: Congressional Research Service.

Kniffel, L. (1996). Criticism follows hoopla at new San Francisco library. *American Libraries, 27*(7), 12–13.

Kochen, M. (Ed.). (1989). *The small world.* Norwood, NJ: Ablex.

Koltsova, O. (2008). Media, state, and responses to globalization in post-communist Russia. In P. Chakravartty & Y. Zhao (Eds.), *Global communications: Toward a transcultural political economy* (pp. 51–74). Lanham, MD: Rowman & Littlefield.

Kopel, D. B., & Olson, J. (1996). Preventing a reign of terror: Civil liberties implications of terrorism legislation. *Oklahoma City University Law Review, 21,* 247–347.

Kranich, N. (2001). Libraries, the Internet, and democracy. In N. Kranich (Ed.), *Libraries and democracy: The cornerstones of liberty* (pp. 383–395). Chicago: American Library Association.

Kretschmar, M., & Morris, M. (1996). Measures of concurrency in networks and the spread of infectious disease. *Mathematical Bioscience, 133,* 165–195.

Kunii, T. L. (2000). Discovering cyberworlds. *IEEE Computer Graphics and Applications,* January/February, 64–65.

Lamont v. Postmaster General. (1965). 381 U.S. 301.

Lancaster, F. W. (Ed). (1993). *Libraries and the future: Essays on the library in the twenty-first century.* New York: Haworth.

Landmark Communications, Inc. v. Virginia. (1978). 435 U.S. 829.

Lankes, R. D., Silverstein, J., & Nicholson, S. (2007). Participatory networks: The library as conversation. *Information Technology and Libraries, 26*(4), 17–33.

Lastra, J. (2000). *Sound technology and the American cinema: Perception, representation, modernity.* New York: Columbia University Press.

Law, D. (2008). Counter culture: Reshaping libraries. *Legal Information Management 8*(1), 11–17.

Learned, W. (1924). *The American public library and the diffusion of knowledge.* New York: Harcourt Brace.

———. (1926). *Libraries and adult education*. Chicago: American Library Association.

Leckie, G. J. (2004). Three perspectives on libraries as public space. *Feliciter, 50*(6), 233–236.

Leckie, G. J., & Buschman, J. E. (Eds). (2009). *Information technology in librarianship: Critical approaches*. Westport, CT: Libraries Unlimited.

Leckie, G. J., & Hopkins, J. (2002). The public place of central libraries: Findings from Toronto and Vancouver. *Library Quarterly, 72*, 326–372.

Lessig, L. (2002). *The future of ideas: The fate of the commons in a connected world*. New York: Vintage.

Leuf, B., & Cunningham, W. (2001). *The wiki way: Quick collaboration on the Web*. Boston: Addison-Wesley.

Levy, L. W. (1966). *Freedom of the press from Zenger to Jefferson: Early American libertarian theories*. Indianapolis: Bobbs-Merrill.

Levy, S. (1984). *Hackers: Heroes of the computer revolution*. Garden City, NY: Anchor Press/Doubleday.

Lewis, J. E. (2008). Defining the nation: 1790–1898. In D. Farber (Ed.), *Security v. liberty: Conflicts between civil liberties and national security in American history* (pp. 117–164). New York: Russell Sage Foundation.

Library of Congress Working Group (2008). *On the record: Report of the Library of Congress Working Group on the Future of Bibliographic Control*. Washington, DC: Library of Congress. Available: http://www.loc.gov/bibliographic-future/

Licklider, J. C. R. (1965). *Libraries of the future*. Cambridge, MA: Massachusetts Institute of Technology Press.

Liesener, J. W. (1983). *Learning at risk: School library media programs in an information world*. Washington, DC: Office of Educational Research and Improvement.

Lindell, M. K., & Perry, R. W. (1992). *Behavioral foundations of community emergency planning*. Washington, DC: Taylor & Francis.

Ling, R. (2007). Informal social capital and ICTs. In B. Anderson, M. Brynin, J. Gershung, & Y. Raban (Eds.), *Information and communication technologies in society: E-living in a digital Europe* (pp. 150–162). London: Routledge.

Lombardo, N. T., Mower, A., & McFarland M. M. (2008). Putting wikis to work in libraries. *Medical Reference Services Quarterly 27*(2), 129–145.

Lu, Y. (2007). The human in human information acquisition: Understanding gatekeeping and proposing new directions in scholarship. *Library & Information Science Research, 29*, 103–123.

Lyons, C. (2007). The library: A distinct local voice? *First Monday, 12*(3). Available: http://www.uic.edu/htbin/cgiwrap/bin/ojs/index.php/fm/article/view/1629/1544

Lyons, D. (2009, June 1). They might be a little evil: Why Google faces antitrust scrutiny. *Newsweek*. Available: http://www.newsweek.com

Lyytinen, K., & King, J. L. (2004). Nothing at the center?: Academic legitimacy in the information systems field. *Journal of the Association of Information Systems, 5*(6), 220–246.

Madison, J. (1791, December 19). Public opinion. *National Gazette* (Philadelphia).

Madison, J., Hamilton, A., & Jay, J. (1789). *Federalist papers* (I. Kramnic, Ed.). New York: Penguin.

Malaby, T. M. (2006). Coding control: Governance and contingency in the production of online worlds. *First Monday*, Special Issue 7. Available: http://firstmonday.org/htbin/cgiwrap/bin/ojs/index.php/fm/article/view/1613

Mamet, D. (2004). Secret names. *Threepenny Review, 96*(Winter), 6.

Manchester, W. (1993). *A world lit only by fire: The medieval mind and the Renaissance; Portrait of an age*. New York: Little, Brown and Company.

Mandel, C. A., & Wolven, R. (1996). Intellectual access to digital documents: Joining proven principles with new technologies. *Cataloging & Classification Quarterly, 22*(3/4), 25–42.

Manoff, M. (2001). The symbolic meaning of libraries in a digital age. *Portal: Libraries and the Academy, 1,* 371–381.

Margolis, M., Resnik, D., & Wolfe, J. D. (1999). Party competition on the Internet in the United States and Britain. *Harvard International Journal of Press Politics, 4*(4), 24–47.

Markoff, J. (2006, November 12). Entrepreneurs see a web guided by common sense. *New York Times.* Available: http://www.nytimes.com/2006/11/12/business/12web.html?_r=1&scp=1&sq=Markoff%20Web%203.0&st=cse&oref=slogin

———. (2009, February 15). A new Internet? The old one is putting us in jeopardy. *New York Times Week in Review,* 1, 4.

Marks, S. (2000). *The riddle of all constitutions.* Oxford: Oxford University Press.

Mart, S. N. (2003). The right to receive information. *Law Library Journal, 95,* 175–189.

Martin, J. (1992). *Cultures in organizations: Three perspectives.* New York: Oxford University Press.

Masie, E. (2005). CNN newsroom in the midst of Katrina—"rapid deployment . . . content objects . . . learning implications." *Public Library Quarterly, 24*(2), 73–76.

Maxwell, T. A. (2005). Constructing consensus: Homeland security as a symbol of government politics and administration. *Government Information Quarterly, 22,* 152–169.

Mayhew, D. R. (2000). *America's congress: Actions in the public sphere, James Madison through Newt Gingrich.* New Haven, CT: Yale University Press.

McChesney, K. (1984). History of libraries, librarianship, and library education. In A. R. Rogers & K. McChesney (Eds.), *The library in society* (pp. 33–60). Littleton, CO: Libraries Unlimited.

McChesney, R. W. (1993). *Telecommunications, mass media, and democracy: The battle for the control of U.S. broadcasting, 1928–1935.* New York: Oxford University Press.

McClure, C. R., & Jaeger, P. T. (2008a). Government information policy research: Importance, approaches, and realities. *Library & Information Science Research, 30,* 257–264.

———. (2008b). *Public libraries and Internet service roles: Measuring and maximizing Internet services.* Chicago: ALA Editions.

McClure, C. R., Jaeger, P. T., & Bertot, J. C. (2007). The looming infrastructure plateau?: Space, funding, connection speed, and the ability of public libraries to meet the demand for free Internet access. *First Monday, 12*(12). Available: http://www.uic.edu/htbin/cgiwrap/bin/ojs/index.php/fm/article/view/2017/1907

McCreadie, M., & Rice, R. E. (1999a). Trends in analyzing access to information, part I: Cross-disciplinary conceptions of access. *Information Processing and Management, 35,* 45–76.

———. (1999b). Trends in analyzing access to information, part II: Unique and integrating conceptualizations. *Information Processing and Management, 35,* 77–99.

McCrossen, A. (2006). "One more cathedral" or "mere lounging places for bummers?" The cultural politics of leisure and the public library in Gilded Age America. *Libraries & the Cultural Record, 41,* 169–188.

McCullough, D. (2001). *John Adams.* New York: Simon & Schuster.

McElfresh, L. K. (2008). Folksonomies and the future of subject cataloging. *Technicalities, 28*(2), 3–6.

McEntire, D. A. (1997). Reflecting on the weaknesses of the international community during IDNDR: Some implications for research and application. *Disaster Prevention and Management, 6*, 221–233.

———. (2002). Coordinating multi-organisational responses to disaster: Lessons from the March 28, 2000, Fort Worth Tornado. *Disaster Prevention and Management, 11*, 369–379.

McGrath, W. E. (2002). Explanation and prediction: Building a unified theory of librarianship, concept and review. *Library Trends, 50*, 350–369.

McIver, W. J., Birdsall, W. F., & Rasmussen, M. (2003). The Internet and the right to communicate. *First Monday, 8*(2). Available: http://www.firstmonday.org/issues/issue8_12/mciver/

McKechnie, L., & Pettigrew, K. E. (2002). Surveying the use of theory in library and information science research: A disciplinary perspective. *Library Trends, 50*(3), 406–417.

McMullen, H. (2000). *American libraries before 1876.* Westport, CT: Greenwood.

Mehra, B., & Srinivasan, R. (2007). The library-community convergence framework for community action. *Libri, 57*(3), 123–139.

Meyer, H. (1998). *All on fire: William Lloyd Garrison and the abolition of slavery.* New York: St. Martin's Griffin.

Miksa, F. (1991). Library and information science: Two paradigms? In P. Vakkari & B. Cronin (Eds.), *Conceptions of library and information science: Historical, empirical and theoretical perspectives.* London: Taylor and Graham.

Milbank, D. (2009, June 24). Stay tuned for more of 'the Obama show.' *Washington Post*, A2.

Miles, S. A. (1967). An introduction to the vocabulary of information technology. *Technical Communications*, Fall, 20–24.

Milgram, S. (1967). The small world problem. *Psychology Today, 2*, 60–67.

Military Commissions Act, Public Law 109–366.

Miller, C. R., & Shepherd, D. (2004). Blogging as social action: A genre analysis of the weblog. *Into the blogosphere: Rhetoric, community, and culture of weblogs.* Available: http://blog.lib.umn.edu/blogosphere/blogging_as_social_action_a_genre_analysis_of_the_weblog.html

Mokros, H. B. (2008). One iSchool's ideas and identity: Doctoral training and research at Rutgers-SCILS 1959–2007. Paper presented at *2008 iConference.*

Morehead, J. (1999). *Introduction to United States government information sources* (6th ed.). Westport, CT: Libraries Unlimited.

Moynihan, D. P. (1998). *Secrecy.* New Haven, CT: Yale University Press.

Mueller, J. (2006). *Overblown: How politicians and the terrorism industry inflate national security threats, and why we believe them.* New York: Free Press.

Murdock, G., & Golding, P. (1989). Information poverty and political inequality: Citizenship in the age of privatized communications. *Journal of Communication, 39*, 180–195.

Murray, D., Schwartz, J., & Lichter, S. R. (2002). *It ain't necessarily so: How the media remake our picture of reality.* New York: Penguin.

Musman, K. (1993). *Technological innovations in libraries, 1860–1960.* Westport, CT: Greenwood.

Nardi, B. A., Schiano, D. J., & Gumbrecht, M. (2004). Blogging as social activity, or would you let 900 million people read your diary? *Proceedings of the Conference on Computer-Supported Cooperative Work* (pp. 222–231). New York: ACM Press.

Naugle, D. K. (2002). *Worldview: The history of a concept.* Grand Rapids, MI: W. B. Eerdmans.

Nerone, J. (1994). *Violence against the press: Policing the public sphere in U.S. history.* New York: Oxford University Press.

Neville, A., & Datray, T. (1993). Planning for equal intellectual access for blind and low vision users. *Library Hi Tech, 11*(1), 67–71.

Nichols, J., & McChesney, R. W. (2005). *Tragedy and farce: How the American media sell wars, spin elections, and destroy democracy.* New York: New Press.

Nunberg, G. (1996). Farwell to the information age. In G. Nunberg (Ed.), *The future of the book* (pp. 103–138). Berkeley: University of California Press.

OCLC. (2007). *Sharing, privacy and trust in our networked world.* Dublin, OH: Author.

———. (2008). *WorldCat at a glance.* Available: http://www.oclc.org/us/en/worldcat/about/default.htm

Office of the Attorney General. (2001). *Memorandum for heads of all federal departments and agencies: The Freedom of Information Act.* Washington, DC: U.S. Department of Justice. Available: http://www.usdoj.gov/oip/011012.htm

Olsen, S. (2004). Google adds major libraries to its database. *CNET News.* Available: http://news.cnet.com/Google-adds-major-libraries-to-its-database/2100–1025_3–5489921.html?tag=nefd.top

Owyang, J. (2008, November 3). *Snapshot of the presidential candidate social networking stats: Nov 3, 2008.* Available: http://www.web-strategist.com/blog/2008/11/03/snapshot-of-presidential-candidate-social-networking-stats-nov-2–2008/

Palen, L., Hiltz, S. R., & Liu, S. (2007). Online forums supporting grassroots participation in emergency preparedness and response. *Communications of the ACM, 50*(3), 54–58.

Pawley, C. (2005). History in the library and information science curriculum: Outline of a debate. *Libraries & the Cultural Record, 40*(3), 223–238.

Paz, O. (2004). The media spectacle comes to Mexico. In T. Hayden (Ed.), *The Zapatista reader* (pp. 30–32). New York: Nation Books.

Pease, D. E., Jr. (2003). The global homeland state: Bush's biopolitical settlement. *Boundary 2, 30*(3), 1–18.

Pelfrey, W. V. (2005). The cycle of preparedness: Establishing a framework to prepare for terrorist threats. *Journal of Homeland Security and Emergency Management, 2*(1), 1–21.

Pettigrew, K. E., & Durrance, J. C. (2000). KALIPER study identifies trends in library and information science education. In D. Bogart & J. C. Blixrud (Eds.), *The Bowker annual library and book trade almanac* (45th ed.). New Providence, NJ: Bowker.

Pew Internet & American Life Project. (2009). *Trend data.* Available: http://www.pewinternet.org/Static-Pages/Data-Tools/Download-Data/Trend-Data.aspx

Pittman, R. (2001). Sex, democracy, and videotape. In N. Kranich (Ed.), *Libraries and democracy: The cornerstones of liberty* (pp. 113–118). Chicago: American Library Association.

Pitts, J., & Stripling, B. (1990). Information power challenge: To provide intellectual and physical access. *School Library Media Quarterly* (Spring 1990), 133–134.

Pool, I. d. S. (1990). *Technologies with boundaries.* Cambridge, MA: Harvard University Press.

Portsea, L. J. (1992). Disaster relief or relief disaster?: A challenge to the international community. *Disasters, 16*, 1–8.

Poster, M. (1997). Cyberdemocracy: Internet and the public sphere. In D. Porter (ed.), *Internet culture* (pp. 201–218). New York: Routledge.

Powell, R. R. (1995). Research competence for PhD students in library and information science. *Journal of Education for Library and Information Science, 36*(4), 319–329.

Powell, R. R., Baker, L. M., & Mika, J. J. (2002). Library and information science practitioners and research. *Library & Information Science Research, 24*, 49–72.

Powell, W. W., & DiMaggio, P. (1991). *The new institutionalism in organizational analysis*. Chicago: University of Chicago Press.

Preer, J. L. (2006). "Louder please": Using historical research to foster professional identity in LIS students. *Libraries & the Cultural Record, 41*, 487–496.

———. (2008). Promoting citizenship: How librarians helped get out the vote in the 1952 presidential election. *Libraries & the Cultural Record, 43*, 1–28.

Price, M. E. (1995). *Television: the public sphere and national identity*. Oxford: Clarendon.

Prior, M. (2007). *Post-broadcast democracy: How media choice increases inequality in political involvement and polarizes elections*. New York: Cambridge University Press.

Project for Excellence in Journalism. (2005). *The state of American media, 2005*. Washington, DC: Author.

Public Agenda. (2006). *Long overdue: A fresh look at public and leadership attitudes about libraries in the 21st Century*. New York: Author.

Puddington, A. (2000). *Broadcasting freedom: The Cold War through triumph of Radio Free Europe and Radio Liberty*. Lexington: University of Kentucky Press.

Pungitore, V. L. (1995). *Innovation and the library: The adoption of new ideas in public libraries*. Westport, CT: Greenwood.

Quinn, A. C. (2003). Keeping the citizenry informed: Early congressional printing and 21st century information policy. *Government Information Quarterly, 20*, 281–293.

Raber, D. (2003). Librarians as organic intellectuals: A Gramscian approach to blind spots and tunnel vision. *Library Quarterly, 73*, 33–53.

Radford, G. P. (2003). Trapped in our own discursive formations: Toward an archaeology of library and information science. *Library Quarterly, 73*, 1–18.

Rankin, V. (1992). Pre-search: Intellectual access to information. *School Library Journal*, March, 168–170.

Raven, J. (2007). Social libraries and library societies in eighteenth century North America. In T. Augst & K. Carpenter (Eds.), *Institutions of reading: The social life of libraries in the United States* (pp. 1–23). Amherst: University of Massachusetts.

Ravid, G., & Rafaeli, S. (2004). Asynchronous discussion groups as small world and scale free networks. *First Monday, 9*(9). Available: http://firstmonday.org/issues/issue9_9/ravid/index

Rawls, J. (1996). *Political liberalism*. New York: Columbia University Press.

Rayport, J. F., & Sviokla, J. J. (1995). Exploiting the virtual value chain. *Harvard Business Review, 6*, 75–85.

Rayward, W. B., & Jenkins, C. (2007). Libraries in times of war, revolution, and social change. *Library Trends, 55*(3), 361–369.

Reed, M. (n.d.). Flame warriors. Available: http://redwing.hutman.net/~mreed/

Reith, D. (1984). The library as social agency. In A. R. Rogers & K. McChesney (Eds.), *The library in society* (pp. 5–16). Littleton, CO: Libraries Unlimited.

Rennison, E. (1994). Galaxy of news: An approach to visualizing and understanding expansive news landscapes. *Proceeding of the 7th annual ACM symposium on user interface software and technology* (pp. 3–12). Marina del Rey, CA: ACM.

Reylea, H. C. (2008). Federal government information policy and public policy analysis: A brief overview. *Library & Information Science Research, 30*, 2–21.

Relyea, H. C., & Halchin, L. E. (2003). Homeland security and information management. In D. Bogart (Ed.), *The Bowker annual: Library and trade almanac 2003* (pp. 231–250). Medford, NJ: Information Today.

Rheingold, H. (1993). *The virtual community*. Reading, MA: Addison-Wesley.

Rich, M. (2009, May 11). Print books are target of pirates on the web. *New York Times*, A1, B4.

Robbins, L. S. (1996). *Censorship and the American library: The American Library Association's response to threats to intellectual freedom*. Westport, CT: Greenwood.

———. *The dismissal of Miss Ruth Brown: Civil rights, censorship, and the American library*. Norman: University of Oklahoma Press.

———. (2007). Responses to the resurrection of Miss Ruth Brown: An essay on the reception of a historical case study. *Libraries & the Cultural Record, 42*, 422–437.

Roberts, A. (2006). *Blacked out: Government secrecy in the information age*. New York: Cambridge University Press.

Roesler, M., & Hawkins, D. T. (1994). Intelligent agents: Software servants for an electronic information world (and more)! *Online, 18*(4), 18–32.

Rogers, R. (2004). *Information politics on the Web*. Cambridge, MA: Massachusetts Institute of Technology Press.

Rogers, R. A. (1984). An introduction to philosophies of librarianship. In A. R. Rogers & K. McChesney (Eds.), *The library in society* (pp. 17–32). Littleton, CO: Libraries Unlimited.

Ross, A., & Caidi, N. (2005). Action and reaction: Libraries in the post 9/11 environment. *Library & Information Science Research, 27*, 97–114.

Rothbauer, P. (2007). Locating the library as place among lesbian, gay, bisexual, and queer patrons. In J. Buschman & G. J. Leckie (Eds.), *The library as place: History, community, and culture*. Westport, CT: Libraries Unlimited.

Sadun, E. (2008). Google copyright deal moves forward. *Ars Technica*. Available: http://arstechnica.com/old/content/2008/11/google-copyright-deal-moves-forward.ars

Samek, T. (2001). *Intellectual freedom and social responsibility in American librarianship, 1967–1974*. Jefferson, NC: McFarland.

Saracevic, T. (1999). Information science. *Journal of the American Society for Information Science, 50*, 1051–1063.

Sarikakis, K. (2008). Regulating the consciousness industry in the European Union: Legitimacy, identity, and the changing state. In P. Chakravartty & Y. Zhao (Eds.), *Global communications: Toward a transcultural political economy* (pp. 51–74). Lanham, MD: Rowman & Littlefield.

Saslow, E. (2008, June 30). Obama rumors fly in Flag City. *Washington Post*. Available: http://www.msnbc.com

Savolainen, R. (2008). *Everyday information practices: A social phenomenological perspective*. Lanham, MD: Rowan & Littlefield.

Schattschneider, E. E. (1969). *Two hundred million Americans in search of a government*. New York: Holt, Rinehart and Winston.

Schutz, A., & Luckmann, T. (1973). *The structures of the life-world*. Evanston, IL: Northwestern University Press.

Seko, S. (2005). Google sued over print library project. *Ars Technica*. Available: http://arstechnica.com/old/content/2005/09/5334.ars

Seymour, W. N., Jr. (1980). *The changing roles of public libraries*. Metuchen, NJ: Scarecrow.

Shane, S., & Mazzetti, M. (2009, April 22). Origins of 'torture' tactics overlooked. *New York Times*. Available: http://www.nytimes.com

Shannon, C. E., & Weaver, W. (1964). *The mathematical theory of communication*. Urbana: University of Illinois Press.

Shera, J. H. (1949). *Foundations of the public library: Origins of public library movement in New England 1629–1855*. Chicago: University of Chicago Press.

———. (1964). Automation and the reference librarian. *Reference Quarterly, 3*(July), 3–7.

———. (1970). *The sociological foundations of librarianship*. New York: Asia Publishing House.

Sherif, M. (1935). A study of some social factors in perception. *Archives of Psychology*, No. 187.

Shim, S-F. (2008). Regional crisis, personal solutions: The media's role in securing neoliberal hegemony in Singapore. In P. Chakravartty & Y. Zhao (Eds.), *Global communications: Toward a transcultural political economy* (pp. 51–74). Lanham, MD: Rowman & Littlefield.

Shneiderman, B. (2008). Science 2.0. *Science, 319*(March 7), 1349–1350.

Shuman, B. A. (2001). *Issues for libraries and information science in the Internet age*. Englewood, CO: Libraries Unlimited.

Simoncelli, T., & Stanley, J. (2005). *Science under siege: The Bush administration's assault on academic and scientific inquiry*. New York: American Civil Liberties Union.

Smith, E. S. (1995). Equal information access and the evolution of American democracy. *Journal of Educational Media & Library Sciences, 33*(2), 158–171.

Smith, G. W. (2004). *The politics of deceit: Saving freedom and democracy from extinction*. Hoboken, NJ: Wiley & Sons.

Smith, R. J., & Eggen, D. (2009, March 3). Bush-era anti-terrorism documents made public. *Washington Post*, A5.

Smolla, R. A. (1992). *Free speech in an open society*. New York: Vintage.

Solomon, W. S. (1993). The contours of media history. In W. S. Solomon & R. W. McChesney (Eds.), *Ruthless criticism: New perspectives in U. S. communication history* (pp. 1–6). Minneapolis: University of Minnesota Press.

Spiteri, L. F. (2006). The use of folksonomies in public library catalogues. *The Serials Librarian, 51*(2), 75–89.

Srinivasan, R. (2006a). Where information society and community voice intersect. *Information Society, 22*, 355–365.

———. (2006b). Indigenous, ethnic and cultural articulations of new media. *International Journal of Cultural Studies, 9*, 497–518.

———. (2007). Ethnomethodological architectures: Information systems driven by cultural and community visions. *Journal of the American Society for Information Science and Technology, 58*, 723–733.

Stanley v. Georgia. (1969). 394 U.S. 557.

Star, S. L., Bowker, G. C., & Bishop, A. P. (2003). Transparency beyond the individual level of scale: Convergence between information artifacts and communities of practice. In A. P. Bishop, N. A. Van House, & B. P. Buttenfield (Eds.), *Digital library use: Social practice in design and evaluation*. Cambridge, MA: Massachusetts Institute of Technology Press.

Star, S. L., & Griesemer, J. R. (1989). Institutional ecology, 'translations' and boundary objects: Amateurs and professionals in Berkeley's Museum of Vertebrate Zoology. *Social Studies of Science, 19*, 387–420.

Starr, P. (2004). *The creation of the media: Political origins of modern communication*. New York: Basic Books.

Stephens, M. (2007). Web 2.0 and libraries, part 2: Trends and technologies. *Library Technology Reports, 43*(5), 10–14.

Stielow, F. (2001). Reconsidering 'arsenals of a democratic culture': Balancing symbol and practice. In N. Kranich (Ed.), *Libraries and democracy: The cornerstones of liberty* (pp. 3–14). Chicago: American Library Association.

Stone, G. R. (2004). *Perilous times: Free speech in wartime from the Sedition Acts of 1798 to the war on terrorism*. New York: Norton.

———. (2008). The Vietnam War: Spying on Americans. In D. Farber (Ed.), *Security v. liberty: Conflicts between civil liberties and national security in American history* (pp. 95–116). New York: Russell Sage Foundation.

Streitz, N., Magerkurth, C., Prante, T., & Rocker, C. (2005). From information design to experience design: Smart artifacts and the disappearing computer. *Interactions, 12*(4), 21–25.

Strickland, L. S. (2003). Copyright's digital dilemma today: Fair use or unfair constraints? Part 1: The battle over file sharing. *Bulletin of the American Society for Information Science and Technology, 30*(1), 7–11.

———. (2004). Copyright's digital dilemma today: Fair use or unfair constraints? Part 2: The DCMA, the TEACH Act, and e-copying restrictions. *Bulletin of the American Society for Information Science and Technology, 30*(2), 18–23.

———. (2005). The information gulag: Rethinking openness in times of national danger. *Government Information Quarterly, 22*, 546–572.

Stvilia, B., Twidale, M. B., Smith, L. C., & Gasser, L. (2008). Information quality work organization in Wikipedia. *Journal of the American Society for Information Science and Technology, 59*(6), 983–1001.

Sullivan, K. (2001). Freedom of expression in the United States: Past and present. In T. R. Hensley (Ed.), *The boundaries of freedom of expression & order in American democracy.* Kent, OH: Kent State University Press.

Sunstein, C. R. (2001). *Republic.com.* Princeton, NJ: Princeton University Press.

———. (2002). The law of group polarization. *Journal of Political Philosophy, 10*, 175–195.

———. (2005). *Laws of fear: Beyond the precautionary principle.* New York: Cambridge University Press.

———. (2008). Neither Hayek nor Habermas. *Public Choice, 134*, 87–95.

Suskind, R. (2004). *The price of loyalty: George W. Bush, the White House, and the education of Paul O'Neill.* New York: Simon & Schuster.

Sutton, S. S. (1998). The panda syndrome II: Innovation, discontinuous change, and LIS education. *Journal of Education for Library and Information Science, 40*, 247–262.

Svenonius, E. (2000). *The intellectual foundation of information organization.* Cambridge, MA: Massachusetts Institute of Technology Press.

Swanson, D. R. (1979). Libraries and the growth of knowledge. *Library Quarterly, 49*, 3–25.

———. (1980). Evolution, libraries, and national information policy. *Library Quarterly, 50*, 76–93.

Tatelman, T. B. (2008). Congress' contempt power: Three mechanisms for enforcing subpoenas. *Government Information Quarterly, 25*, 592–624.

Taylor, A. G. (1999). *The organization of information.* Westport, CT: Libraries Unlimited.

Taylor, A. G. & Joudrey, D. N. (2009). *The organization of information* (3rd ed.). Westport, CT: Libraries Unlimited.

Taylor, R. S. (1966). Professional aspects of information science and technology. In C. A. Cuadra (Ed.), *Annual review of information science and technology, volume 1.* New York: John Wiley & Sons.

Taylor, T. L. (2006). Beyond management: Considering participatory design and governance in player culture. *First Monday*, Special Issue 7. Available: http://firstmonday.org/htbin/cgiwrap/bin/ojs/index.php/fm/article/view/1611/1526

Thom, A. (1967). *Megalithic sites in Britain.* New York: Oxford University Press.

Thompson, B. (2006, August 13). Search me? Google wants to digitize every book, publishers say read the fine print first. *Washington Post*, D1, D7.

Thompson, K. M. (2006). *Multidisciplinary approaches to information poverty and their implications for information access.* Unpublished dissertation, Florida State University.

———. (2008). Remembering Elfreda Chatman: A champion of theory development in library and information science education. *Journal of Education for Library and Information Science, 50*, 119–126.

Thompson, K. M., McClure, C. R., & Jaeger, P. T. (2003). Evaluating federal websites: Improving e-government for the people. In J. F. George (Ed.), *Computers in society: Privacy, ethics, and the Internet* (pp. 400–412). Upper Saddle River, NJ: Prentice Hall.

Tierney, K. J. (2006). Social inequality, hazards, and disasters. In R. J. Daniels, D. F. Kettl, & H. Kunreuther (Eds.), *On risk and disaster: Lessons from Hurricane Katrina* (pp. 109–128). Philadelphia: University of Pennsylvania Press.

Tierney, K. J., Lindell, M. K., & Perry, R. W. (2001). *Facing the unexpected: Disaster preparedness and response in the United States.* Washington, DC: Joseph Henry Press.

Tisdale, S. (1997). Silence, please: The public library as entertainment center. *Harper's Magazine,* March, 65–73.

Toulmin, S. (1990). *Cosmopolis: The hidden agenda of modernity.* Chicago: University of Chicago Press.

Travers, J., & Milgram, S. (1969). An experimental study of the small world problem. *Sociometry, 32*(4), 425–443.

Travis, H. (2006). Building universal digital libraries: An agenda for copyright reform. *Pepperdine Law Review, 33,* 761–833.

Truman, D. (1965). *The governmental process: Political interest and public opinion.* New York: Knopf.

———. (1971). *The governmental process: Political interests and public opinion* (2nd ed.). New York: Knopf.

Turner, J. C. (1987). *Rediscovering the social group: A self-categorization theory.* New York: Blackwell.

United Nations. (1986). *New information technologies and development.* New York: UN Center for Science and Technology Development.

Uniting and Strengthening America by Providing Appropriate Tools Required to Intercept and Obstruct Terrorism (USA PATRIOT) Act of 2001, Public Law 107–56.

Usher, N. (2008). Reviewing fauxtography: A blog-driven challenge to mass media power without the promises of networked publicity. *First Monday, 13*(12). Available: http://firstmonday.org/htbin/cgiwrap/bin/ojs/index.php/fm/article/view/2158/2055

Vaidhyanathan, S. (2004). *The anarchist in the library: How the clash between freedom and control is hacking the real world and crashing the system.* New York: Basic Books.

———. (2006, December 28). Me, 'person of the year'? No thanks. *MSNBC.* Available: http://www.msnbc.msn.com/id/16371425/

Van House, N. A., & Sutton, S. (1996). The panda syndrome: An ecology of LIS education. *Journal of Education for Library and Information Science, 37*(2), 131–147.

Van Sant, W. (2009, June 8). Librarians now add social work to their resumes. *St. Petersburg Times.* Available: http://www.tampabay.com/

Van Slyck, A. A. (1995). *Free to all: Carnegie libraries and American culture, 1890–1920.* Chicago: University of Chicago Press.

Vargas, J. A. (2008a, December 28). Politics is no longer local: It's viral. *Washington Post,* B1.

———. (2008b, November 20). Obama raised half a billion online. *Washington Post.* Available: http://voices.washingtonpost.com/44/2008/11/20/obama_raised_half_a_billion_on.html

Viegas, F. B. (2005). Bloggers' expectations of privacy and accountability: An initial survey. *Journal of Computer-Mediated Communication, 10*(3), article 12. Available: http://jcmc.indiana.edu/vol10/issue3/viegas.html

Walsh, J. P. (1988). Selectivity and selective perception: An investigation of managers' belief structures and information processing. *Academy of Management Journal, 31,* 873–889.

Wang, I-W. (2008). Schoolhouse rock is no longer enough: The presidential signing statements controversy and its implications for library professionals. *Law Library Journal, 100*(4), 619–637.

Warner, M. (1993). The public sphere and the cultural mediation of print. In W. S. Solomon & R. W. McChesney (Eds.), *Ruthless criticism: New perspectives in U. S. communication history* (pp. 7–37). Minneapolis: University of Minnesota Press.

Wasserman, H. (2006). New media in a new democracy: An exploration of the potential of the Internet for civil society groups in South Africa. In K. Sarikakis, & D. K. Thussu (Eds.), *Ideologies of the Internet* (pp. 299–316). Cresskill, NJ: Hampton.

Watts, D. J. (1999). *Small worlds: The dynamics of networks between order and randomness.* Princeton, NJ: Princeton University Press.

Watts, D. J., & Strogatz, S. H. (1998). Collective dynamics of 'small-world' networks. *Science, 393,* 440–442.

Waugh, W. L., & Sylves, R. T. (2002). Organizing the war on terrorism. *Public Administration Review, 62,* 145–153.

Weber, M. (1978). *Economy and society.* Berkeley: University of California Press.

Webster, F. (2002). *Theories of the information society.* London: Routledge.

Weech, T. L., & Pluzhenskaia, M. (2005). LIS education and multidisciplinarity: An exploratory study. *Journal of Education for Library and Information Science, 36*(4), 319–329.

Wei, C. (2004). Formation of norms in a blog community. *Into the blogosphere: Rhetoric, community, and culture of weblogs.* Available: http://blog.lib.umn.edu/blogosphere/formation_of_norms.html

Weimann, G. (1996). Cable news comes to the holy land: The impact of cable TV on Israeli viewers. *Journal of Broadcasting & Electronic Media, 40*(2), 243–257.

Wells, A. T., & Rainie, L. (2008). The Internet as social ally. *First Monday, 13*(11). Available: http://www.uic.edu/htbin/cgiwrap/bin/ojs/index.php/fm/article/view/2198/2051

West, T. G. (1997). *Vindicating the founders: Race, sex, class, and justice in the origins of America.* Lanham, MD: Rowman & Littlefield.

White House Office, Chief of Staff. (2002). *Action to safeguard information relating to weapons of mass destruction and other sensitive documents related to homeland security.* Washington: Author.

White House Office. (2003). Executive Order 13292. 68 *Federal Register* 15315.

White House. (2006). *The federal response to Hurricane Katrina: Lessons learned.* Washington, DC: Author.

Wiegand, W. A. (1976). *The politics of an emerging profession: The American Library Association, 1876–1917.* New York: Greenwood.

———. (1989). *An active instrument for propaganda: The American public library during World War I.* Westport, CT: Greenwood.

———. (1996). *Irrepressible reformer: A biography of Melvil Dewey.* Chicago: American Library Association.

———. (1999). Tunnel vision and blind spots: What the past tells us about present; reflections on the twentieth-century history of American librarianship. *Library Quarterly, 69,* 1–32.

———. (2005). Critiquing the curriculum. *American Libraries, 36*(1), 58–61.

Wikipedia. (2006a). Wikipedia. Available: http://en.wikipedia.org/wiki/Wikipedia

———. (2006b). Congressional staffer edits to Wikipedia. Available: http://en.wikipedia.org

———. (2006c). Lamest edit wars. Available: http://en.wikipedia.org/wiki/Lamest_edit_wars_ever

———. (2008). Six degrees of Kevin Bacon. Available: http://en.wikipedia.org/wiki/Six_Degrees_of_Kevin_Bacon

———. (2009). Folksonomy. Available: http://en.wikipedia.org/wiki/Folksonomy

Wilbur, S. P. (2000). An archaeology of cyberspaces: Virtuality, community, identity. In D. Bell & B. M. Kennedy, (Eds.), *The cybercultures reader* (pp. 45–55). London and New York: Routledge.

Wilhelm, A. G. (2000). *Democracy in the digital age: Challenges to political life in cyberspace*. London: Routledge.

Will, B. H. (2001). The public library as community crisis center. *Library Journal, 126*(20), 75–77.

Williams, R. (1958/1989). Culture is ordinary. In R. Gable (Ed.), *Resources of hope: Culture, democracy, socialism* (pp. 3–18). London: Verso.

———. (1968/1989). The idea of a common culture. In R. Gable (Ed.), *Resources of hope: Culture, democracy, socialism* (pp. 32–38). London: Verso.

Wilson, P. (1983). *Second-hand knowledge: An inquiry into cognitive authority*. Westport, CT: Greenwood.

———. (1996). Interdisciplinary research and information overload. *Library Trends, 45*(2), 192–203.

Wilson, T. D. (1999). Models in information behavior research. *Journal of Documentation, 55*(3), 249–270.

———. (2000). Human information behavior. *Information Science 3*(2), 49–56.

Winchester, S. (2003). *Krakatoa: The day the world exploded*. New York: Harper Perennial.

Wise, C. R. (2006). Organizing for homeland security after Katrina: Is adaptive management what's missing? *Public Administration Review, 66*, 302–318.

Wood, G. (2004). Academic original sin: Plagiarism, the Internet, and libraries. *Journal of Academic Librarianship, 30*, 237–242.

Wood, J. (1992). *History of international broadcasting*. London: Peregrinus.

———. (2000). *History of international broadcasting, volume 2*. London: Institute of Electrical Engineers.

Woods, B. (1993). *Communication, technology and the development of people*. London: Routledge.

Yoo, J. (2008). FDR, civil liberties, and the war on terror. In D. Farber (Ed.), *Security v. liberty: Conflicts between civil liberties and national security in American history* (pp. 42–66). New York: Russell Sage Foundation.

Zaret, D. (2000). *Origins of democratic culture: Printing, petitions, and the public sphere in early-modern England*. Princeton, NJ: Princeton University Press.

Zhao, Y. (2008). Neoliberal strategies, socialist legacies: Communication and state transformation in China. In P. Chakravartty & Y. Zhao (Eds.), *Global communications: Toward a transcultural political economy* (pp. 51–74). Lanham, MD: Rowman & Littlefield.

Zittrain, J., & Palfrey, J. (2008). Internet filtering: The politics and mechanisms of control. In R. Deibert, J. Palfrey, R. Rohozinski, & J. Zittrain (Eds.), *Access denied: The practice and policy of global Internet filtering* (pp. 29–56). Cambridge, MA: Massachusetts Institute of Technology Press.

Index